Modern
Differential Calculus

Modern Differential Calculus

Satish Gupta

Oxford Book Company

Jaipur ▪ New Delhi

Published By

Oxford Book Company

Regd. Off. "Bony Residency", Gate No. 2
Opp. Tilak Public School, Vishweswariya Nagar,
Gopalpura Road, Jaipur - 302018, India
Ph: 0141 - 2761280, Telefax: 2761381
email - oxfordbookjpr@gmail.com
website - oxfordbookcompany.com

Delhi Office: 102, 1st Floor, Satyam House,
4327/3, Ansari Road, Darya Ganj,
New Delhi - 110002, India
Ph: 011-45652440

Distributed By
KEYA INTERNATIONAL PRESS LLP
E-mail: keyainternationalpressllp@gmail.com

First published in India, 2026

ISBN : 978-93-5524-767-4

Typeset by Shivangi Computers, Jaipur

Printed and bound in India by Thomson Press Ltd.,
New Delhi

Contents

Preface

The field of calculus has remained an indispensable pillar in the edifice of modern mathematics and applied sciences. Among its branches, differential calculus stands as a foundational discipline that bridges the realms of pure mathematics, engineering, physics, economics, and countless other domains. The evolution of differential calculus has not only revolutionized the way we perceive change and motion but has also enabled groundbreaking advancements in diverse fields. Modern Differential Calculus is crafted with the objective of providing a lucid, comprehensive, and contemporary approach to this profound subject while maintaining mathematical rigor and conceptual depth.

This book is designed to serve as a valuable resource for undergraduate and postgraduate students, educators, and researchers alike. It offers a structured exploration of fundamental principles, beginning with the core concepts of limits, continuity, and differentiability, progressing towards advanced topics such as higher-order derivatives, partial differentiation, and their extensive applications. The presentation of the subject matter is carefully curated to ensure clarity, logical progression, and an intuitive understanding of complex concepts.

A distinguishing feature of this book is its emphasis on real-world applications, demonstrating how differential calculus operates as a powerful analytical tool in various scientific and engineering disciplines. Each chapter is enriched with illustrative examples, problem sets, and detailed explanations that facilitate both theoretical comprehension and practical problem-solving skills. Furthermore, significant attention has been given to modern computational techniques and their role in the contemporary mathematical landscape.

In an era where mathematical computation is becoming increasingly digital, this book also incorporates discussions on numerical differentiation, algorithmic approaches, and the use of computational tools to enhance learning and application. Special care has been taken to present historical insights, enabling readers to appreciate the contributions of great mathematicians who have shaped the development of differential calculus over centuries.

This book will not only serve as a guide for academic excellence but also ignite intellectual curiosity and a deeper appreciation for the elegance of calculus. Any constructive feedback and suggestions for improvement are most welcome, as the pursuit of mathematical knowledge is a continuous journey of refinement and discovery.

— ***Satish Gupta***

Chapter 1

Introduction

In mathematics, differential calculus is a subfield of calculus concerned with the study of the rates at which quantities change. It is one of the two traditional divisions of calculus, the other being integral calculus.

The primary objects of study in differential calculus are the derivative of a function, related notions such as the differential, and their applications. The derivative of a function at a chosen input value describes the rate of change of the function near that input value.

The process of finding a derivative is called differentiation. Geometrically, the derivative at a point equals the slope of the tangent line to the graph of the function at that point. For a real-valued function of a single real variable, the derivative of a function at a point generally determines the best linear approximation to the function at that point.

Differential calculus and integral calculus are connected by the fundamental theorem of calculus, which states that differentiation is the reverse process to integration.

Differentiation has applications to nearly all quantitative disciplines. For example, in physics, the derivative of the displacement of a moving body with respect to time is the velocity of the body, and the derivative of velocity with respect to time is acceleration. Newton's second law of motion states that the derivative of the momentum of a body equals the force applied to the body.

The reaction rate of a chemical reaction is a derivative. In operations research, derivatives determine the most efficient ways to transport materials and design factories. Derivatives are frequently used to find the maxima and minima of a function. Equations involving derivatives are called differential equations and are fundamental in describing natural phenomena. Derivatives and their generalizations appear in many fields of mathematics, such as complex analysis, functional analysis, differential geometry, measure theory and abstract algebra.

The Derivative

Suppose that x and y are real numbers and that y is a function of x, that is, for every value of x, there is a corresponding value of y. This relationship is written as $y = f(x)$. If $f(x)$ is the equation for a straight line, then there are two real numbers m and b such that $y = m\ x + b$. m is called the slope and can be determined from the formula:

$$m = \frac{\text{change in } y}{\text{change in } x} = \frac{\Delta y}{\Delta x},$$

where the symbol Δ (the uppercase form of the Greek letter Delta) is an abbreviation for "change in". It follows that $\Delta y = m\ \Delta x$.

A general function is not a line, so it does not have a slope. The derivative of f at the point x is the best possible approximation to the idea of the slope of f at the point x.

It is usually denoted $f'(x)$ or dy/dx. Together with the value of f at x, the derivative of f determines the best linear approximation, or linearization, of f near the point x. This latter property is usually taken as the definition of the derivative.

A closely related notion is the differential of a function.

When x and y are real variables, the derivative of f at x is the slope of the tangent line to the graph of f at x. Because the source and target of f are one-dimensional, the derivative of f is a real number.

If x and y are vectors, then the best linear approximation to the graph of f depends on how f changes in several directions at once. Taking the best linear approximation in a single direction determines a partial derivative, which is usually denoted $\partial y/\partial x$. The linearization of f in all directions at once is called the total derivative.

History of Differentiation

The concept of a derivative in the sense of a tangent line is a very old one, familiar to Greek geometers such as Euclid (c. 300 BC), Archimedes (c. 287–212 BC) and Apollonius of Perga (c. 262–190 BC). Archimedes also introduced the use of infinitesimals, although these were primarily used to study areas and volumes rather than derivatives and tangents.

The use of infinitesimals to study rates of change can be found in Indian mathematics, perhaps as early as 500 AD, when the astronomer and mathematician Aryabhata (476–550) used infinitesimals to study the motion of the moon.

The use of infinitesimals to compute rates of change was developed significantly by Bhâskara II (1114–1185); indeed, it has been argued that many of the key notions of differential calculus can be found in his work, such as "Rolle's theorem". The Persian mathematician, Sharaf al-Dîn al-Tûsî (1135–1213), was the first to discover the derivative of cubic polynomials, an important result in differential calculus; his *Treatise on Equations* developed concepts related to differential calculus, such as the derivative function and the maxima and minima of curves, in order to solve cubic equations which may not have positive solutions.

The modern development of calculus is usually credited to Isaac Newton (1643–1727) and Gottfried Leibniz (1646–1716), who provided independent and unified approaches to differentiation and derivatives. The key insight, however, that earned them this credit, was the fundamental theorem of calculus relating differentiation and integration: this rendered obsolete most previous methods for computing areas and volumes, which had not been significantly extended since the time of Ibn al-Haytham (Alhazen).

For their ideas on derivatives, both Newton and Leibniz built on significant earlier work by mathematicians such as Isaac Barrow (1630–1677), René Descartes (1596–1650), Christiaan Huygens (1629–1695), Blaise Pascal (1623–1662) and John Wallis (1616–1703). Isaac Barrow is generally given credit for the early development of the derivative. Nevertheless, Newton and Leibniz remain key figures in the history of differentiation, not least because Newton was the first to apply differentiation to theoretical physics, while Leibniz systematically developed much of the notation still used today.

Since the 17th century many mathematicians have contributed to the theory of differentiation. In the 19th century, calculus was put on a much more rigorous footing by mathematicians such as Augustin Louis Cauchy (1789–1857), Bernhard Riemann (1826–1866), and Karl Weierstrass (1815–1897). It was also during this period that the differentiation was generalized to Euclidean space and the complex plane.

Applications of Derivatives

Optimization: If f is a differentiable function on R (or an open interval) and x is a local maximum or a local minimum of f, then the derivative of f at x is zero; points where $f'(x) = 0$ are called *critical points* or *stationary points* (and the value of f at x is called a *critical value*). (The definition of a critical point is sometimes extended to include points where the derivative does not exist.) Conversely, a critical point x of f can be analysed by considering the second derivative of f at x:

- if it is positive, x is a local minimum;
- if it is negative, x is a local maximum;
- if it is zero, then x could be a local minimum, a local maximum, or neither. (For example, $f(x) = x^3$ has a critical point at $x = 0$, but it has neither a maximum nor a minimum there, whereas $f(x) = \pm x^4$ has a critical point at $x = 0$ and a minimum and a maximum, respectively, there.)

This is called the second derivative test. An alternative approach, called the first derivative test, involves considering the sign of the f' on each side of the critical point.

Taking derivatives and solving for critical points is therefore often a simple way to find local minima or maxima, which can be useful in optimization. By the extreme value theorem, a continuous function on a closed interval must attain its minimum and maximum values at least once. If the function is differentiable, the minima and maxima can only occur at critical points or endpoints.

This also has applications in graph sketching: once the local minima and maxima of a differentiable function have been found, a rough plot of the graph can be obtained from the observation that it will be either increasing or decreasing between critical points.

In higher dimensions, a critical point of a scalar valued function is a point at which the gradient is zero. The second derivative test can still be used to analyse critical points by considering the eigenvalues of the Hessian matrix of second partial derivatives of the function at the critical point. If all of the eigenvalues are positive, then the point is a local minimum; if all are negative, it is a local maximum. If there are some positive and some negative eigenvalues, then the critical point is a saddle point, and if none of these cases hold (i.e., some of the eigenvalues are zero) then the test is inconclusive.

Calculus of Variations

Calculus of variations is a field of mathematics, or more specifically calculus, that deals with maximizing or minimizing functionals, which are mappings from a set of functions to the real numbers. Functionals are often expressed as definite integrals involving functions and their derivatives. The interest is in *extremal* functions that make the functional attain a maximum or minimum value – or *stationary* functions – those where the rate of change of the functional is zero. Perhaps the simplest example of such a problem is to find the curve of shortest length, or geodesic, connecting two points.

If there are no constraints, the solution is obviously a straight line between the points. However, if the curve is constrained to lie on a surface in space, then the solution is less obvious, and possibly many solutions may exist. Such solutions are known as geodesics. A related problem is posed by Fermat's principle: light follows the path of shortest optical length connecting two points, where the optical length depends upon the material of the medium. One corresponding concept in mechanics is the principle of least action. Many important problems involve functions of several variables. Solutions of boundary value problems for the Laplace equation satisfy the Dirichlet principle. Plateau's problem requires finding a surface of minimal area that spans a given contour in space: the solution or solutions can often be found by dipping a wire frame in a solution of soap suds. Although such experiments are relatively easy to perform, their mathematical interpretation is far from simple: there may be more than one locally minimizing surface, and they may have non-trivial topology.

History

The calculus of variations may be said to begin with the brachistochrone curve problem raised by Johann Bernoulli (1696). It immediately occupied the attention of Jakob Bernoulli and the Marquis de l'Hôpital, but Leonhard Euler first elaborated the subject. His contributions began in 1733, and his Elementa Calculi Variationum gave to the science its name. Lagrange contributed extensively to the theory, and Legendre (1786) laid down a method, not entirely satisfactory, for the discrimination of maxima and minima.

Isaac Newton and Gottfried Leibniz also gave some early attention to the subject. To this discrimination Vincenzo Brunacci (1810), Carl Friedrich Gauss (1829), Siméon Poisson (1831), Mikhail Ostrogradsky (1834), and Carl Jacobi (1837) have been among the contributors. An important general work is that of Sarrus (1842) which was condensed and improved by Cauchy (1844). Other valuable treatises and memoirs have been written by Strauch (1849), Jellett (1850), Otto Hesse (1857), Alfred Clebsch (1858), and Carll (1885), but perhaps the most important work of the century is that of Weierstrass. His celebrated course on the theory is epoch-making, and it may be asserted that he was the first to place it on a firm and unquestionable foundation.

The 20th and the 23rd Hilbert problems published in 1900 encouraged further development. In the 20th century David Hilbert, Emmy Noether, Leonida Tonelli, Henri Lebesgue and Jacques Hadamard among others made significant contributions. Marston Morse applied calculus of variations in what is now called Morse theory. Lev Pontryagin, Ralph Rockafellar and Clarke developed new mathematical tools for optimal control theory, a generalisation of calculus of variations.

Euler–Lagrange Equation

Under ideal conditions, the maxima and minima of a given function may be located by finding the points where its derivative vanishes (i.e., is equal to zero). By analogy, solutions of smooth variational problems may be obtained by solving the associated Euler–Lagrange equation.

Consider the functional

$$A[f] = \int_{x_1}^{x_2} L[x, f(x), f'(x)]dx$$

where $f'(x) = \dfrac{df}{dx}$ and where x_1 and x_2 are constants.

The function f should have at least one derivative in order to satisfy the requirements for valid application of the function; further, if the functional $A[f]$ attains its local minimum at f_0 and $\eta(x)$ is an arbitrary function that has at least one derivative and vanishes at the endpoints x_1 and x_2, then we must have

$$A[f_0] \leq A[f_0 + \epsilon\eta]$$

for any number ε close to 0. Therefore, with $f = f_0 + \epsilon\eta$ the first variation of A must vanish,

$$\left.\frac{dA[f_0 + \epsilon\eta]}{d\epsilon}\right|_{\epsilon=0} = \int_{x_1}^{x_2} \left.\frac{dL}{d\epsilon}\right|_{\epsilon=0} dx = 0\,.$$

Since L is a function of f and f', using the chain rule,

$$\frac{dL}{d\epsilon} = \frac{\partial L}{\partial f}\frac{df}{d\epsilon} + \frac{\partial L}{\partial f'}\frac{df'}{d\epsilon}$$

and since $\dfrac{df}{d\epsilon} = \eta$ and $\dfrac{df'}{d\epsilon} = \eta'$,

$$\frac{dL}{d\epsilon} = \frac{\partial L}{\partial f}\eta + \frac{\partial L}{\partial f'}\eta'.$$

The last term in the second line vanishes because $\eta \equiv 0$ at the end points. As previously mentioned the left side of the equation is zero, so according to the fundamental lemma of calculus of variations, the part of the integrand in parentheses is zero and we find that L will satisfy

$$\frac{\partial L}{\partial f} - \frac{d}{dx}\frac{\partial L}{\partial f'} = 0,$$

which is called the Euler–Lagrange equation.

In general this gives a second-order ordinary differential equation which can be solved to obtain the extremal *f*. The Euler–Lagrange equation is a necessary, but not sufficient, condition for an extremal. Sufficient conditions for an extremal are discussed in the references.

In order to illustrate this process, consider the problem of finding the shortest curve in the plane that connects two points (x_1, y_1) and (x_2, y_2). The arc length is given by

$$A[f] = \int_{x_1}^{x_2} \sqrt{1+[f'(x)]^2}\, dx,$$

with

$$f'(x) = \frac{df}{dx},$$

and where $y = f(x)$, $f(x_1) = y_1$, and $f(x_2) = y_2$. Now, let $f(x) = f_0(x) + \epsilon\eta(x)$, where $f_0(x)$ is a minimizer for $A[f]$ and $\epsilon \in \mathbb{R}$ close to zero. Then

$$\frac{d}{d\epsilon}\int_{x_1}^{x_2} \sqrt{1+[f_{0'}(x)+\epsilon\eta'(x)]^2}\,dx\bigg|_{\epsilon=0} = \int_{x_1}^{x_2} \frac{(f_{0'}(x)+\epsilon\eta'(x))\eta'(x)}{\sqrt{1+[f_{0'}(x)+\epsilon\eta'(x)]^2}}\Bigg|_{\epsilon=0}$$

$$dx = \int_{x_1}^{x_2} \frac{f_{0'}(x)\eta'(x)}{\sqrt{1+[f_{0'}(x)]^2}}dx = 0,$$

for any choice of the function η(though for the next step we will need to require that $\eta(x)$ vanishes at x_1, x_2). We may interpret this condition as the vanishing of all directional derivatives of $A[f_0]$ in the space of differentiable functions, and this is formalized by requiring the Fréchet derivative of *A* to vanish at f_0. If we assume that f_0 has two continuous derivatives (or if we consider weak derivatives), then we may use integration by parts:

$$\int_a^b u(x)\eta'(x)dx = [u(x)\eta(x)]_a^b - \int_a^b u'(x)\eta(x)dx$$

with the substitution

$$u(x) = \frac{f_{0'}(x)}{\sqrt{1+[f_{0'}(x)]^2}}$$

then we have

$$[u(x)\eta(x)]_{x_1}^{x_2} - \int_{x_1}^{x_2} \eta(x)\frac{d}{dx}\left[\frac{f_{0'}(x)}{\sqrt{1+[f_{0'}(x)]^2}}\right]dx = 0,$$

but the first term is zero since $\eta(x)$ was chosen to vanish at x_1 and x_2 where the evaluation is taken. Therefore,

$$\int_{x_1}^{x_2} \eta(x)\frac{d}{dx}\left[\frac{f_{0'}(x)}{\sqrt{1+[f_{0'}(x)]^2}}\right]dx = 0$$

for any twice differentiable function η that vanishes at the endpoints of the interval.

We can now apply the fundamental lemma of calculus of variations: If

$$I = \int_{x_1}^{x_2} \eta(x)H(x)dx = 0$$

for any sufficiently differentiable function $\eta(x)$ within the integration range that vanishes at the endpoints of the interval, then it follows that $H(x)$ is identically zero on its domain.

Therefore,

$$\frac{d}{dx}\left[\frac{f_{0'}(x)}{\sqrt{1+[f_{0'}(x)]^2}}\right] = 0.$$

It follows from this equation that

$$\frac{d^2 f_0}{dx^2} = 0 \Rightarrow f_0(x) = c_1 x + c_2$$

and hence the extremals are straight lines.

Beltrami Identity

In physics problems it frequently turns out that $\partial L / \partial x = 0$. In that case, the Euler-Lagrange equation can be simplified using the Beltrami identity:

$$L - f' \frac{\partial L}{\partial f'} = C, \text{ [1]}$$

where C is a constant. The left hand side is the Legendre transformation of L with respect to f'.

Du Bois Reymond's Theorem

The discussion thus far has assumed that extremal functions possess two continuous derivatives, although the existence of the integral A requires only first derivatives of trial functions. The condition that the first variation vanish at an extremal may be regarded as a weak form of the Euler-Lagrange equation.

The theorem of du Bois Reymond asserts that this weak form implies the strong form. If L has continuous first and second derivatives with respect to all of its arguments, and if

$$\frac{\partial^2 L}{(\partial f')^2} \neq 0,$$

then f_0 has two continuous derivatives, and it satisfies the Euler-Lagrange equation.

Lavrentiev Phenomenon

Hilbert was the first to give good conditions for the Euler Lagrange equations to give a stationary solution. Within a convex area and a positive thrice differentiable Lagrangian the solutions are composed of a countable collection of sections that either go along the boundary or satisfy the Euler Lagrange equations in the interior. However Lavrentiev in 1926 showed that there are circumstances where there is no optimum solution but one can be approached arbitrarily closely by increasing numbers of sections. For instance the following:

$$L(t, x, x') = (x^3 - t)^2 x'^6,$$

$$x(0) = 0, x(1) = 1.$$

Here a zig zag path gives a better solution than any smooth path and increasing the number of sections improves the solution.

Functions of Several Variables

Variational problems that involve multiple integrals arise in numerous applications. For example, if $\varphi(x,y)$ denotes the displacement of a membrane above the domain D in the x,y plane, then its potential energy is proportional to its surface area:

$$U[\varphi] = \iint_D \sqrt{1 + \nabla\varphi\cdot\nabla\varphi}\, dx\, dy.$$

Plateau's problem consists of finding a function that minimizes the surface area while assuming prescribed values on the boundary of D; the solutions are called minimal surfaces. The Euler-Lagrange equation for this problem is nonlinear:

$$\varphi_{xx}(1+\varphi_y^2) + \varphi_{yy}(1+\varphi_x^2) - 2\varphi_x\varphi_y\varphi_{xy} = 0.$$

Dirichlet's Principle

It is often sufficient to consider only small displacements of the membrane, whose energy difference from no displacement is approximated by

$$V[\varphi] = \frac{1}{2}\iint_D \nabla\varphi\cdot\nabla\varphi\, dx\, dy.$$

The functional V is to be minimized among all trial functions φ that assume prescribed values on the boundary of D. If u is the minimizing function and v is an arbitrary smooth function that vanishes on the boundary of D, then the first variation of $V[u+\epsilon v]$ must vanish:

$$\frac{d}{d\epsilon}V[u+\epsilon v]\Big|_{\epsilon=0} = \iint_D \nabla u\cdot\nabla v\, dx\, dy = 0.$$

Provided that u has two derivatives, we may apply the divergence theorem to obtain

$$\iint_D \nabla\cdot(v\nabla u)\, dx\, dy = \iint_D \nabla u\cdot\nabla v + v\nabla\cdot\nabla u\, dx\, dy = \iint_C v\frac{\partial u}{\partial n}\, ds,$$

where C is the boundary of D, s is arclength along C and $\partial u / \partial n$ is the normal derivative of u on C. Since v vanishes on C and the first variation vanishes, the result is

$$\iint_D v\nabla \cdot \nabla u dxdy = 0$$

for all smooth functions v that vanish on the boundary of D. The proof for the case of one dimensional integrals may be adapted to this case to show that

$$\nabla \cdot \nabla u = 0 \text{ in } D.$$

The difficulty with this reasoning is the assumption that the minimizing function u must have two derivatives. Riemann argued that the existence of a smooth minimizing function was assured by the connection with the physical problem: membranes do indeed assume configurations with minimal potential energy. Riemann named this idea the Dirichlet principle in honour of his teacher Dirichlet. However Weierstrass gave an example of a variational problem with no solution: minimize

$$W[\varphi] = \int_{-1}^{1} (x\varphi')^2 dx$$

among all functions φ that satisfy $\varphi(-1) = -1$ and $\varphi(1) = 1$. W can be made arbitrarily small by choosing piecewise linear functions that make a transition between -1 and 1 in a small neighbourhood of the origin. However, there is no function that makes W=0. Eventually it was shown that Dirichlet's principle is valid, but it requires a sophisticated application of the regularity theory for elliptic partial differential equations.

Generalization to Other Boundary Value Problems

A more general expression for the potential energy of a membrane is

$$V[\varphi] = \iint_D \left[\frac{1}{2}\nabla\varphi \cdot \nabla\varphi + f(x,y)\varphi\right] dxdy + \int_C \left[\frac{1}{2}\sigma(s)\varphi^2 + g(s)\varphi\right] ds.$$

This corresponds to an external force density $f(x, y)$ in D, an external force $g(s)$ on the boundary C, and elastic forces with modulus $\sigma(s)$ acting on C. The function that minimizes the potential energy with no restriction on its boundary values will be denoted by u. Provided that f and g are continuous, regularity theory implies that the minimizing function u will have two derivatives. In taking the first variation, no boundary condition need be imposed on the increment v. The first variation of $V[u+\epsilon v]$ is given by

$$\iint_D [\nabla u \cdot \nabla v + fv]\,dx\,dy + \int_C [\sigma uv + gv]\,ds = 0.$$

If we apply the divergence theorem, the result is

$$\iint_D [-v\nabla \cdot \nabla u + vf]\,dx\,dy + \int_C v\left[\frac{\partial u}{\partial n} + \sigma u + g\right] ds = 0.$$

If we first set v=0 on C, the boundary integral vanishes, and we conclude as before that

$$-\nabla \cdot \nabla u + f = 0$$

in D. Then if we allow v to assume arbitrary boundary values, this implies that u must satisfy the boundary condition

$$\frac{\partial u}{\partial n} + \sigma u + g = 0,$$

on C. Note that this boundary condition is a consequence of the minimizing property of u: it is not imposed beforehand. Such conditions are called natural boundary conditions.

The preceding reasoning is not valid if σ vanishes identically on C. In such a case, we could allow a trial function $\varphi \equiv c$, where c is a constant. For such a trial function,

$$V[c] = c\left[\iint_D f\,dx\,dy + \int_C g\,ds\right].$$

By appropriate choice of c, V can assume any value unless the quantity inside the brackets vanishes. Therefore the variational problem is meaningless unless

$$\iint_D f\,dx\,dy + \int_C g\,ds = 0.$$

This condition implies that net external forces on the system are in equilibrium. If these forces are in equilibrium, then the variational problem has a solution, but it is not unique, since an arbitrary constant may be added. Further details and examples are in Courant and Hilbert (1953).

Eigenvalue Problems

Both one-dimensional and multi-dimensional eigenvalue problems can be formulated as variational problems.

Sturm-Liouville Problems

The Sturm-Liouville eigenvalue problem involves a general quadratic form

$$Q[\varphi]=\int_{x_1}^{x_2}\left[p(x)\varphi'(x)^2+q(x)\varphi(x)^2\right]dx,$$

where φ is restricted to functions that satisfy the boundary conditions

$$\varphi(x_1)=0, \quad \varphi(x_2)=0.$$

Let R be a normalization integral

$$R[\varphi]=\int_{x_1}^{x_2} r(x)\varphi(x)^2\,dx.$$

The functions $p(x)$ and $r(x)$ are required to be everywhere positive and bounded away from zero. The primary variational problem is to minimize the ratio Q/R among all φ satisfying the endpoint conditions. It is shown below that the Euler-Lagrange equation for the minimizing u is

$$-(pu')'+qu-\lambda ru=0,$$

where λ is the quotient

$$\lambda=\frac{Q[u]}{R[u]}.$$

It can be shown that the minimizing u has two derivatives and satisfies the Euler-Lagrange equation. The associated λ will be denoted by λ_1; it is the lowest eigenvalue for this equation and boundary conditions. The associated minimizing function will be denoted by $u_1(x)$. This variational characterization of eigenvalues leads to the Rayleigh-Ritz method: choose an approximating u as a linear combination of basis functions (for example trigonometric functions) and carry out a finite-dimensional minimization among such linear combinations. This method is often surprisingly accurate.

The next smallest eigenvalue and eigenfunction can be obtained by minimizing Q under the additional constraint

$$\int_{x_1}^{x_2} r(x)u_1(x)\varphi(x)dx = 0.$$

This procedure can be extended to obtain the complete sequence of eigenvalues and eigenfunctions for the problem.

The variational problem also applies to more general boundary conditions.

Instead of requiring that φ vanish at the endpoints, we may not impose any condition at the endpoints, and set

$$Q[\varphi] = \int_{x_1}^{x_2} \left[p(x)\varphi'(x)^2 + q(x)\varphi(x)^2 \right] dx + a_1\varphi(x_1)^2 + a_2\varphi(x_2)^2,$$

where a_1 and a_2 are arbitrary. If we set $\varphi = u + \epsilon v$ the first variation for the ratio Q/R is

$$V_1 = \frac{2}{R[u]} \left(\begin{array}{l} \int_{x_1}^{x_2} [p(x)u'(x)v'(x) + q(x)u(x)v(x) - \lambda u(x)v(x)] \\ dx + a_1 u(x_1)v(x_1) + a_2 u(x_2)v(x_2) \end{array} \right),$$

where λ is given by the ratio $Q[u]/R[u]$ as previously. After integration by parts,

$$\frac{R[u]}{2}V_1 = \int_{x_1}^{x_2} v(x)\left[-(pu')' + qu - \lambda ru \right] dx + v(x_1)[-p(x_1)u'(x_1)$$
$$+a_1 u(x_1)] + v(x_2)[p(x_2)u'(x_2) + a_2 u(x_2)].$$

If we first require that v vanish at the endpoints, the first variation will vanish for all such v only if

$$-(pu')' + qu - \lambda ru = 0 \quad \text{for} \quad x_1 < x < x_2.$$

If u satisfies this condition, then the first variation will vanish for arbitrary v only if

$$-p(x_1)u'(x_1) + a_1 u(x_1) = 0, \quad \text{and} \quad p(x_2)u'(x_2) + a_2 u(x_2) = 0.$$

These latter conditions are the natural boundary conditions for this problem, since they are not imposed on trial functions for the minimization, but are instead a consequence of the minimization.

Eigenvalue Problems in Several Dimensions

Eigenvalue problems in higher dimensions are defined in analogy with the one-dimensional case. For example, given a domain D with boundary B in three dimensions we may define

$$Q[\varphi] = \iiint_D p(X)\nabla\varphi\cdot\nabla\varphi + q(X)\varphi^2\,dxdydz + \iint_B \sigma(S)\varphi^2\,dS,$$

and $R[\varphi] = \iiint_D r(X)\varphi(X)^2\,dxdydz.$

Let u be the function that minimizes the quotient $Q[\varphi]/R[\varphi]$, with no condition prescribed on the boundary B. The Euler-Lagrange equation satisfied by u is

$$-\nabla\cdot(p(X)\nabla u) + q(x)u - \lambda r(x)u = 0,$$

where

$$\lambda = \frac{Q[u]}{R[u]}.$$

The minimizing u must also satisfy the natural boundary condition

$$p(S)\frac{\partial u}{\partial n} + \sigma(S)u = 0,$$

on the boundary B. This result depends upon the regularity theory for elliptic partial differential equations. Many extensions, including completeness results, asymptotic properties of the eigenvalues and results concerning the nodes of the eigenfunctions are in Courant and Hilbert (1953).

Applications

Some applications of the Calculus of variations include:

- The derivation of the Catenary shape
- The Brachistochrone problem
- Isoperimetric problems
- Geodesics on surfaces
- Minimal surfaces and Plateau's problem
- Optimal Control.

Fermat's Principle

Fermat's principle states that light takes a path that (locally) minimizes the optical length between its endpoints. If the x-coordinate is chosen as the parameter along the path, and $y = f(x)$ along the path, then the optical length is given by

$$A[f] = \int_{x=x_0}^{x_1} n(x, f(x))\sqrt{1 + f'(x)^2}\, dx,$$

where the refractive index $n(x, y)$ depends upon the material.

If we try $f(x) = f_0(x) + \epsilon f_1(x)$ then the first variation of A (the derivative of A with respect to ε) is

$$\delta A[f_0, f_1] = \int_{x=x_0}^{x_1} \left[\frac{n(x, f_0) f_{0'}(x) f_{1'}(x)}{\sqrt{1 + f_{0'}(x)^2}} + n_y(x, f_0) f_1 \sqrt{1 + f_{0'}(x)^2} \right] dx.$$

After integration by parts of the first term within brackets, we obtain the Euler-Lagrange equation

$$-\frac{d}{dx}\left[\frac{n(x, f_0) f_{0'}}{\sqrt{1 + f_{0'}^2}} \right] + n_y(x, f_0)\sqrt{1 + f_{0'}(x)^2} = 0.$$

The light rays may be determined by integrating this equation. This formalism is used in the context of Lagrangian optics and Hamiltonian optics.

Snell's Law

There is a discontinuity of the refractive index when light enters or leaves a lens. Let

$$n(x, y) = n_- \quad \text{if} \quad x < 0,$$

$$n(x, y) = n_+ \quad \text{if} \quad x > 0,$$

where n_- and n_+ are constants. Then the Euler-Lagrange equation holds as before in the region where $x<0$ or $x>0$, and in fact the path is a straight line there, since the refractive index is constant. At the $x=0$, f must be continuous, but f' may be discontinuous. After integration by parts in the separate regions and using the Euler-Lagrange equations, the first variation takes the form

$$\delta A[f_0, f_1] = f_1(0)\left[n_- \frac{f_{0'}(0_-)}{\sqrt{1+f_{0'}(0_-)^2}} - n_+ \frac{f_{0'}(0_+)}{\sqrt{1+f_{0'}(0_+)^2}} \right].$$

The factor multiplying n_- is the sine of angle of the incident ray with the x axis, and the factor multiplying n_+ is the sine of angle of the refracted ray with the x axis. Snell's law for refraction requires that these terms be equal. As this calculation demonstrates, Snell's law is equivalent to vanishing of the first variation of the optical path length.

Fermat's Principle in Three Dimensions

It is expedient to use vector notation: let $X = (x_1, x_2, x_3)$, let t be a parameter, let $X(t)$ be the parametric representation of a curve C, and let $\dot{X}(t)$ be its tangent vector. The optical length of the curve is given by

$$A[C] = \int_{t=t_0}^{t_1} n(X)\sqrt{\dot{X}\cdot\dot{X}}\,dt.$$

Note that this integral is invariant with respect to changes in the parametric representation of C. The Euler-Lagrange equations for a minimizing curve have the symmetric form

$$\frac{d}{dt}P = \sqrt{\dot{X}\cdot\dot{X}}\,\nabla n,$$

where

$$P = \frac{n(X)\dot{X}}{\sqrt{\dot{X}\cdot\dot{X}}}.$$

It follows from the definition that P satisfies

$$P\cdot P = n(X)^2.$$

Therefore the integral may also be written as

$$A[C] = \int_{t=t_0}^{t_1} P\cdot\dot{X}\,dt.$$

This form suggests that if we can find a function ψ whose gradient is given by P, then the integral A is given by the difference of ψ at the endpoints of the interval of integration. Thus the problem of studying the curves that make the integral stationary can be related to the study of the level surfaces of ψ. In order to find such a function, we turn to the wave equation, which governs the propagation of light. This formalism is used in the context of Lagrangian optics and Hamiltonian optics.

Connection with the Wave Equation

The wave equation for an inhomogeneous medium is

$$u_{tt} = c^2 \nabla \cdot \nabla u,$$

where c is the velocity, which generally depends upon X. Wave fronts for light are characteristic surfaces for this partial differential equation: they satisfy

$$\varphi_t^2 = c(X)^2 \nabla \varphi \cdot \nabla \varphi.$$

We may look for solutions in the form

$$\varphi(t, X) = t - \psi(X).$$

In that case, ψ satisfies

$$\nabla \psi \cdot \nabla \psi = n^2,$$

where $n = 1/c$. According to the theory of first-order partial differential equations, if $P = \nabla \psi$, then P satisfies

$$\frac{dP}{ds} = n \nabla n,$$

along a system of curves (the light rays) that are given by

$$\frac{dX}{ds} = P.$$

These equations for solution of a first-order partial differential equation are identical to the Euler-Lagrange equations if we make the identification

$$\frac{ds}{dt} = \frac{\sqrt{\dot{X} \cdot \dot{X}}}{n}.$$

We conclude that the function ψ is the value of the minimizing integral A as a function of the upper end point. That is, when a family of minimizing curves is constructed, the values of the optical length satisfy the characteristic equation corresponding the wave equation. Hence, solving the associated partial differential equation of first order is equivalent to finding families of solutions of the variational problem. This is the essential content of the Hamilton-Jacobi theory, which applies to more general variational problems.

Action Principle

In classical mechanics, the action, S, is defined as the time integral of the Lagrangian, L. The Lagrangian is the difference of energies,

$$L = T - U,$$

where T is the kinetic energy of a mechanical system and U its potential energy. Hamilton's principle (or the action principle) states that the motion of a conservative holonomic (integrable constraints) mechanical system is such that the action integral

$$S = \int_{t=t_0}^{t_1} L(x, \dot{x}, t)dt$$

is stationary with respect to variations in the path $x(t)$. The Euler-Lagrange equations for this system are known as Lagrange's equations:

$$\frac{d}{dt}\frac{\partial L}{\partial \dot{x}} = \frac{\partial L}{\partial x},$$

and they are equivalent to Newton's equations of motion (for such systems). The conjugate momenta P are defined by

$$p = \frac{\partial L}{\partial \dot{x}}.$$

For example, if

$$T = \frac{1}{2}m\dot{x}^2,$$

then

$$p = m\dot{x}.$$

Hamiltonian mechanics results if the conjugate momenta are introduced in place of x, and the Lagrangian L is replaced by the Hamiltonian H defined by

$$H(x, p, t) = p\dot{x} - L(x, \dot{x}, t)$$

The Hamiltonian is the total energy of the system: $H = T + U$. Analogy with Fermat's principle suggests that solutions of Lagrange's equations (the particle trajectories) may be described in terms of level surfaces of some function of X. This function is a solution of the Hamilton-Jacobi equation:

$$\frac{\partial \psi}{\partial t} + H\left(x, \frac{\partial \psi}{\partial x}, t\right) = 0.$$

Physics

Calculus is of vital importance in physics: many physical processes are described by equations involving derivatives, called differential equations.

Physics is particularly concerned with the way quantities change and evolve over time, and the concept of the "time derivative" — the rate of change over time — is essential for the precise definition of several important concepts. In particular, the time derivatives of an object's position are significant in Newtonian physics:

- velocity is the derivative (with respect to time) of an object's displacement (distance from the original position)
- acceleration is the derivative (with respect to time) of an object's velocity, that is, the second derivative (with respect to time) of an object's position.

For example, if an object's position on a line is given by

$$x(t) = -16t^2 + 16t + 32,$$

then the object's velocity is

$$\dot{x}(t) = x'(t) = -32t + 16,$$

and the object's acceleration is

$$\ddot{x}(t) = x''(t) = -32,$$

which is constant.

Chapter 2

Overview of Differential Calculus

Calculus (Latin, *calculus*, a small stone used for counting) is a branch of mathematics focused on limits, functions, derivatives, integrals, and infinite series. This subject constitutes a major part of modern mathematics education. It has two major branches, differential calculus and integral calculus, which are related by the fundamental theorem of calculus.

Calculus is the study of change, in the same way that geometry is the study of shape and algebra is the study of operations and their application to solving equations. A course in calculus is a gateway to other, more advanced courses in mathematics devoted to the study of functions and limits, broadly called mathematical analysis. Calculus has widespread applications in science, economics, and engineering and can solve many problems for which algebra alone is insufficient.

Calculus has historically been called "the calculus of infinitesimals", or "infinitesimal calculus". More generally, *calculus* (plural *calculi*) refers to any method or system of calculation guided by the symbolic manipulation of expressions. Some examples of other well-known calculi are propositional calculus, variational calculus, lambda calculus, pi calculus, and join calculus.

History of Calculus

Calculus, historically known as infinitesimal calculus, is a mathematical discipline focused on limits, functions, derivatives, integrals, and infinite series. Ideas leading up to the notions of function, derivative, and integral were developed throughout the 17th century, but the decisive step was made by Isaac Newton and Gottfried Leibniz. Publication of Newton's main treatises took many years, whereas Leibniz published first (*Nova methodus*, 1684) and the whole subject was subsequently marred by a priority dispute between the two inventors of calculus.

Ancient Greek Precursors of the Calculus

Greek mathematicians are credited with a significant use of infinitesimals. Democritus is the first person recorded to consider seriously the division of objects into an infinite number of cross-sections, but his inability to rationalize discrete cross-sections with a cone's smooth slope prevented him from accepting the idea. At approximately the same time, Zeno of Elea discredited infinitesimals further by his articulation of the paradoxes which they create.

Antiphon and later Eudoxus are generally credited with implementing the method of exhaustion, which made it possible to compute the area and volume of regions and solids by breaking them up into an infinite number of recognizable shapes.

Archimedes of Syracuse developed this method further, while also inventing heuristic methods which resemble modern day concepts somewhat. It was not until the time of Newton that these methods were incorporated into a general framework of integral calculus. It should not be thought that infinitesimals were put on a rigorous footing during this time, however. Only when it was supplemented by a proper geometric proof would Greek mathematicians accept a proposition as true.

Archimedes was the first to find the tangent to a curve, other than a circle, in a method akin to differential calculus. While studying the spiral, he separated a point's motion into two components, one radial motion component and one circular motion component, and then continued to add the two component motions together, thereby finding the tangent to the curve. The pioneers of the calculus such as Isaac Barrow and Johann Bernoulli were diligent students of Archimedes.

Medieval Development

The mathematical study of continuity was revived in the fourteenth century by the Oxford Calculators and French collaborators such as Nicole Oresme. They proved the "Merton mean speed theorem": that a uniformly accelerated body travels the same distance as a body with uniform speed whose speed is half the final velocity of the accelerated body.

Pioneers of Modern Calculus

In the 17th century, European mathematicians Isaac Barrow, René Descartes, Pierre de Fermat, Blaise Pascal, John Wallis and others discussed the idea of a derivative. In particular, in *Methodus ad disquirendam maximam et minima* and in *De tangentibus linearum curvarum*, Fermat developed an adequality method for determining maxima, minima, and tangents to various curves that was equivalent to differentiation. Isaac Newton would later write that his own early ideas about calculus came directly from "Fermat's way of drawing tangents."

On the integral side, Cavalieri developed his method of indivisibles in the 1630s and 40s, providing a more modern form of the ancient Greek method of exhaustion, and computing Cavalieri's quadrature formula, the area under the curves x^n of higher degree, which had previously only been computed for the parabola, by Archimedes.

Torricelli extended this work to other curves such as the cycloid, and then the formula was generalized to fractional and negative powers by Wallis in 1656.

In a 1659 treatise, Fermat is credited with an ingenious trick for evaluating the integral of any power function directly. Fermat also obtained a technique for finding the centres of gravity of various plane and solid figures, which influenced further work in quadrature.

James Gregory, influenced by Fermat's contributions both to tangency and to quadrature, was then able to prove a restricted version of the second fundamental theorem of calculus in the mid-17th century.

The first full proof of the fundamental theorem of calculus was given by Isaac Barrow. Newton and Leibniz, building on this work, independently developed the surrounding theory of infinitesimal calculus in the late 17th century. Also, Leibniz did a great deal of work with developing consistent and useful notation and concepts. Newton provided some of the most important applications to physics, especially of integral calculus.

The first proof of Rolle's theorem was given by Michel Rolle in 1691 using methods developed by the Dutch mathematician Johann van Waveren Hudde.

The mean value theorem in its modern form was stated by Bernard Bolzano and Augustin-Louis Cauchy (1789–1857) also after the founding of modern calculus. Important contributions were also made by Barrow, Huygens, and many others.

Newton and Leibniz

Before Newton and Leibniz, the word "calculus" was a general term used to refer to any body of mathematics, but in the following years, "calculus" became a popular term for a field of mathematics based upon their insights. The purpose of this section is to examine Newton and Leibniz's investigations into the developing field of infinitesimal calculus. Specific importance will be put on the justification and descriptive terms which they used in an attempt to understand calculus as they themselves conceived it.

By the middle of the seventeenth century, European mathematics had changed its primary repository of knowledge. In comparison to the last century which maintained Hellenistic mathematics as the starting point for research, Newton, Leibniz and their contemporaries increasingly looked towards the works of more modern thinkers. Europe had become home to a burgeoning mathematical community and with the advent of enhanced institutional and organizational bases a new level of organization and academic integration was being achieved. Importantly, however, the community lacked formalism; instead it consisted of a disordered mass of various methods, techniques, notations, theories, and paradoxes.

Newton came to calculus as part of his investigations in physics and geometry. He viewed calculus as the scientific description of the generation of motion and magnitudes. In comparison, Leibniz focused on the tangent problem and came to believe that calculus was a metaphysical explanation of change. These differences in approach should neither be overemphasized nor underappreciated. Importantly, the core of their insight was the formalization of the inverse properties between the integral and the differential. This insight had been anticipated by their predecessors, but they were the first to conceive calculus as a system in which new rhetoric and descriptive terms were created. Their unique discoveries lay not only in their imagination, but also in their ability to synthesize the insights around them into a universal algorithmic process, thereby forming a new mathematical system.

Newton

Newton completed no definitive publication formalizing his Fluxional Calculus; rather, many of his mathematical discoveries were transmitted through correspondence, smaller papers or as embedded aspects in his other definitive compilations, such as the *Principia* and *Opticks*. Newton would begin his mathematical training as the chosen heir of Isaac Barrow in Cambridge. His incredible aptitude was recognized early and he quickly learned the current theories. By 1664 Newton had made his first important contribution by advancing the binomial theorem, which he had extended to include fractional and negative exponents. Newton succeeded in expanding the applicability of the binomial theorem by applying the algebra of finite quantities in an analysis of infinite series. He showed a willingness to view infinite series not only as approximate devices, but also as alternative forms of expressing a term.

Many of Newton's critical insights occurred during the plague years of 1665-1666 which he later described as, "the prime of my age for invention and minded mathematics and [natural] philosophy more than at any time since." It was during his plague-induced isolation that the first written conception of Fluxionary Calculus was recorded in the unpublished *De Analysi per Aequationes Numero Terminorum Infinitas*. In this paper, Newton determined the area under a curve by first calculating a momentary rate of change and then extrapolating the total area. He began by reasoning about an indefinitely small triangle whose area is a function of x and y.

He then reasoned that the infinitesimal increase in the abscissa will create a new formula where $x = x + o$ (importantly, o is the letter, not the digit 0). He then recalculated the area with the aid of the binomial theorem, removed all quantities containing the letter o and re-formed an algebraic expression for the area. Significantly, Newton would then "blot out" the quantities containing *o* because terms "multiplied by it will be nothing in respect to the rest".

At this point Newton had begun to realize the central property of inversion. He had created an expression for the area under a curve by considering a momentary increase at a point. In effect, the fundamental theorem of calculus was built into his calculations. While his new formulation offered incredible potential, Newton was well aware of its logical limitations at the time. He admits that "errors are not to be disregarded in mathematics, no matter how small" and that what he had achieved was "shortly explained rather than accurately demonstrated."

In an effort to give calculus a more rigorous explication and framework, Newton compiled in 1671 the *Methodus Fluxionum et Serierum Infinitarum*. In this book, Newton's strict empiricism shaped and defined his Fluxional Calculus. He exploited instantaneous motion and infinitesimals informally. He used math as a methodological tool to explain the physical world. The base of Newton's revised Calculus became continuity; as such he redefined his calculations in terms of continual flowing motion. For Newton, variable magnitudes are not aggregates of infinitesimal elements, but are generated by the indisputable fact of motion.

Newton attempted to avoid the use of the infinitesimal by forming calculations based on ratios of changes. In the Methodus Fluxionum he defined the rate of generated change as a fluxion, which he represented by a dotted letter, and the quantity generated he defined as a fluent. For example, if x and y are fluents, then x and y are their respective fluxions.

This revised calculus of ratios continued to be developed and was maturely stated in the 1676 text *De Quadratura Curvarum* where Newton came to define the present day derivative as the ultimate ratio of change, which he defined as the ratio between evanescent increments (the ratio of fluxions) purely at the moment in question. Essentially, the ultimate ratio is the ratio as the increments vanish into nothingness. Importantly, Newton explained the existence of the ultimate ratio by appealing to motion;

"For by the ultimate velocity is meant that, with which the body is moved, neither before it arrives at its last place, when the motion ceases nor after but at the very instant when it arrives... the ultimate ratio of evanescent quantities is to be understood, the ratio of quantities not before they vanish, not after, but with which they vanish" Newton developed his Fluxional Calculus in an attempt to evade the informal use of infinitesimals in his calculations.

Leibniz

While Newton began development of his fluxional calculus in 1665-1666 his findings did not become widely circulated until later. In the intervening years Leibniz also strove to create his calculus. In comparison to Newton who came to math at an early age, Leibniz began his rigorous math studies with a mature intellect. He was a polymath, and his intellectual interests and achievements involved metaphysics, law, economics, politics, logic, and mathematics. In order to understand Leibniz's reasoning in calculus his background should be kept in mind. Particularly, his metaphysics which considered the world as an infinite aggregate of indivisible monads. and his plans of creating a precise formal logic whereby, "a general method in which all truths of the reason would be reduced to a kind of calculation."

In 1672 Leibniz met the mathematician Huygens who convinced Leibniz to dedicate significant time to the study of mathematics. By 1673 he had progressed to reading Pascal's *Traité des Sinus du Quarte Cercle* and it was during his largely autodidactic research that Leibniz said a light turned on. Leibniz, like Newton, saw the tangent as a ratio but declared it as simply the ratio between ordinates and abscissas. He continued to argue that the integral was in fact the sum of the ordinates for infinitesimal intervals in the abscissa, in effect, a sum of an infinite number of rectangles. From these definitions the inverse relationship became clear and Leibniz quickly realized the potential to form a whole new system of mathematics. Where Newton shied away from the use of infinitesimals, Leibniz made it the cornerstone of his notation and calculus.

In the manuscripts of 25 October to 11 November 1675, Leibniz recorded his discoveries and experiments with various forms of notation. He was acutely aware of the notational terms used and his earlier plans to form a precise logical symbolism became evident. Eventually, Leibniz denoted the infinitesimal increments of abscissas and ordinates dx and dy, and the summation of infinitely many infinitesimally thin rectangles as a long s (ſ), which became the present integral symbol $\int$.

Importantly, while Leibniz's notation is used by modern mathematics, his logical base was different than our current one. Leibniz embraced infinitesimals and wrote extensively so as, "not to make of the infinitely small a mystery, as had Pascal." Towards this end he defined them "not as a simple and absolute zero, but as a relative zero... that is, as an evanescent quantity which yet retains the character of that which is disappearing." Alternatively, he defines them as, "less than any given quantity." For Leibniz, the world was an aggregate of infinitesimal points and the lack of scientific proof for their existence did not trouble him. Infinitesimals to Leibniz were ideal quantities of a different type from appreciable numbers. The truth of continuity was proven by existence itself. For Leibniz the principle of continuity and thus the validity of his Calculus was assured. Three hundred years after Leibniz's work, Abraham Robinson showed that using infinitesimal quantities in calculus could be given a solid foundation.

Legacy

The rise of Calculus stands out as a unique moment in mathematics. Calculus is the math of motion and change, and as such, its invention required the creation of a new mathematical system. Importantly, Newton and Leibniz did not create the same Calculus and they did not conceive of modern Calculus. While they were both involved in the process of creating a mathematical system to deal with variable quantities their elementary base was different. For Newton, change was a variable quantity over time and for Leibniz it was the difference ranging over a sequence of infinitely close values. Notably, the descriptive terms each system created to describe change was different.

Historically, there was much debate over whether it was Newton or Leibniz who first "invented" calculus. This argument, the Leibniz and Newton calculus controversy, involving Leibniz, who was German, and the Englishman Newton, led to a rift in the European mathematical community lasting over a century. Leibniz was the first to publish his investigations; however, it is well established that Newton had started his work several years prior to Leibniz and had already developed a theory of tangents by the time Leibniz became interested in the question. Much of the controversy centres on the question whether Leibniz had seen certain early manuscripts of Newton before publishing his own memoirs on the subject. Newton began his work on calculus no later than 1666, and Leibniz did not begin his work until 1673. Leibniz visited England in 1673 and again in 1676, and was shown some of Newton's unpublished writings. He also corresponded with several English scientists (as well as with Newton himself), and may have gained access to Newton's manuscripts through them. It is not known how much this may have influenced Leibniz. The initial accusations were made by students and supporters of the two great scientists at the turn of the century, but after 1711 both of them became personally involved, accusing each other of plagiarism.

The priority dispute had an effect of separating English-speaking mathematicians from those in the continental Europe for many years. Only in the 1820s, due to the efforts of the Analytical Society, did Leibnizian analytical calculus became accepted in England. Today, both Newton and Leibniz are given credit for independently developing the basics of calculus. It is Leibniz, however, who is credited with giving the new discipline the name it is known by today: "calculus". Newton's name for it was "the science of fluents and fluxions".

The work of both Newton and Leibniz is reflected in the notation used today. Newton introduced the notation f for the derivative of a function f. Leibniz introduced the symbol $\int$ for the integral and wrote the derivative of a function y of the variable x as $\frac{dy}{dx}$, both of which are still in use.

Integrals

Niels Henrik Abel seems to have been the first to consider in a general way the question as to what differential expressions can be integrated in a finite form by the aid of ordinary functions, an investigation extended by Liouville. Cauchy early undertook the general theory of determining definite integrals, and the subject has been prominent during the 19th century. Frullani's theorem (1821), Bierens de Haan's work on the theory (1862) and his elaborate tables (1867), Dirichlet's lectures (1858) embodied in Meyer's treatise (1871), and numerous memoirs of Legendre, Poisson, Plana, Raabe, Sohncke, Schlömilch, Elliott, Leudesdorf, and Kronecker are among the noteworthy contributions.

Eulerian integrals were first studied by Euler and afterwards investigated by Legendre, by whom they were classed as Eulerian integrals of the first and second species, as follows:

$$\int_0^1 x^{n-1}(1-x)^{n-1}\,dx$$

$$\int_0^\infty e^{-x}x^{n-1}\,dx$$

although these were not the exact forms of Euler's study.

If n is an integer, it follows that:

$$\int_0^\infty e^{-x}x^{n-1}dx = (n-1)!,$$

but the integral converges for all positive real n and defines an analytic continuation of the factorial function to all of the complex plane except for poles at zero and the negative integers. To it Legendre assigned the symbol Γ, and it is now called the gamma function. Besides being analytic over the positive reals, Γ also enjoys the uniquely defining property that $\log\Gamma$ is convex, which aesthetically justifies this analytic continuation of the factorial function over any other analytic continuation. To the subject Dirichlet has contributed an important theorem (Liouville, 1839), which has been elaborated by Liouville, Catalan, Leslie Ellis, and others. On the evaluation of $\Gamma(x)$ and $\log\Gamma(x)$ Raabe (1843–44), Bauer (1859), and Gudermann (1845) have written. Legendre's great table appeared in 1816.

Symbolic Methods

Symbolic methods may be traced back to Taylor, and the much debated analogy between successive differentiation and ordinary exponentials had been observed by numerous writers before the nineteenth century. Arbogast (1800) was the first, however, to separate the symbol of operation from that of quantity in a differential equation. François (1812) and Servois (1814) seem to have been the first to give correct rules on the subject. Hargreave (1848) applied these methods in his memoir on differential equations, and Boole freely employed them. Grassmann and Hermann Hankel made great use of the theory, the former in studying equations, the latter in his theory of complex numbers.

Calculus of Variations

The calculus of variations may be said to begin with a problem of Johann Bernoulli's (1696). It immediately occupied the attention of Jakob Bernoulli and the Marquis de l'Hôpital, but Euler first elaborated the subject. His contributions began in 1733, and his *Elementa Calculi Variationum* gave to the science its name. Lagrange contributed extensively to the theory, and Legendre (1786) laid down a method, not entirely satisfactory, for the discrimination of maxima and minima. To this discrimination Brunacci (1810), Gauss (1829), Poisson (1831), Ostrogradsky (1834), and Jacobi (1837) have been among the contributors. An important general work is that of Sarrus (1842) which was condensed and improved by Cauchy (1844). Other valuable treatises and memoirs have been written by Strauch (1849), Jellett (1850), Hesse (1857), Clebsch (1858), and Carll (1885), but perhaps the most important work of the century is that of Weierstrass. His course on the theory may be asserted that he was the first to place calculus on a firm and rigorous foundation.

Applications

The application of the infinitesimal calculus to problems in physics and astronomy was contemporary with the origin of the science. All through the eighteenth century these applications were multiplied, until at its close Laplace and Lagrange had brought the whole range of the study of forces into the realm of analysis. To Lagrange (1773) we owe the introduction of the theory of the potential into dynamics, although the name "potential function" and the fundamental memoir of the subject are due to Green (1827, printed in 1828).

The name "potential" is due to Gauss (1840), and the distinction between potential and potential function to Clausius. With its development are connected the names of Dirichlet, Riemann, von Neumann, Heine, Kronecker, Lipschitz, Christoffel, Kirchhoff, Beltrami, and many of the leading physicists of the century.

It is impossible in this place to enter into the great variety of other applications of analysis to physical problems. Among them are the investigations of Euler on vibrating chords; Sophie Germain on elastic membranes; Poisson, Lamé, Saint-Venant, and Clebsch on the elasticity of three-dimensional bodies; Fourier on heat diffusion; Fresnel on light; Maxwell, Helmholtz, and Hertz on electricity; Hansen, Hill, and Gyldén on astronomy; Maxwell on spherical harmonics; Lord Rayleigh on acoustics; and the contributions of Dirichlet, Weber, Kirchhoff, F. Neumann, Lord Kelvin, Clausius, Bjerknes, MacCullagh, and Fuhrmann to physics in general.

The labors of Helmholtz should be especially mentioned, since he contributed to the theories of dynamics, electricity, etc., and brought his great analytical powers to bear on the fundamental axioms of mechanics as well as on those of pure mathematics.

Furthermore, infinitesimal calculus was introduced into the social sciences, starting with Neoclassical economics. Today, it is a valuable tool in mainstream economics.

Islamic Mathematics

In the 11th century, when Ibn al-Haytham (known as *Alhacen* in Europe), an Iraqi mathematician working in Egypt, devised what is now known as "Alhazen's problem", which leads to an equation of the fourth degree, in his *Book of Optics*. While solving this problem, he was the first mathematician to derive the formula for the sum of the fourth powers, using a method that is readily generalizable for determining the general formula for the sum of any integral powers. He performed an integration in order to find the volume of a paraboloid, and was able to generalize his result for the integrals of polynomials up to the fourth degree. He thus came close to finding a general formula for the integrals of polynomials, but he was not concerned with any polynomials higher than the fourth degree.

Foundations

In calculus, *foundations* refers to the rigorous development of a subject from precise axioms and definitions. In early calculus the use of infinitesimal quantities was thought unrigorous, and was fiercely criticized by a number of authors, most notably Michel Rolle and Bishop Berkeley. Berkeley famously described infinitesimals as the ghosts of departed quantities in his book *The Analyst* in 1734. A recent study argues that Leibnizian calculus was more solidly grounded than Berkeley's empiricist critique thereof. Working out a rigorous foundation for calculus occupied mathematicians for much of the century following Newton and Leibniz and is still to some extent an active area of research today.

Several mathematicians, including Maclaurin, attempted to prove the soundness of using infinitesimals, but it would not be until 150 years later when, due to the work of Cauchy and Weierstrass, a means was finally found to avoid mere "notions" of infinitely small quantities.

The foundations of differential and integral calculus had been laid. In Cauchy's writing, we find a versatile spectrum of foundational approaches, including a definition of continuity in terms of infinitesimals, and a (somewhat imprecise) prototype of an (ε, δ)-definition of limit in the definition of differentiation.

In his work Weierstrass formalized the concept of limit and eliminated infinitesimals. Following the work of Weierstrass, it eventually became common to base calculus on limits instead of infinitesimal quantities. Bernhard Riemann used these ideas to give a precise definition of the integral. It was also during this period that the ideas of calculus were generalized to Euclidean space and the complex plane.

In modern mathematics, the foundations of calculus are included in the field of real analysis, which contains full definitions and proofs of the theorems of calculus. The reach of calculus has also been greatly extended. Henri Lebesgue invented measure theory and used it to define integrals of all but the most pathological functions. Laurent Schwartz introduced Distributions, which can be used to take the derivative of any function whatsoever. Limits are not the only rigorous approach to the foundation of calculus. An alternative is Abraham Robinson's non-standard analysis. Robinson's approach, developed in the 1960s, uses technical machinery from mathematical logic to augment the real number system with infinitesimal and infinite numbers, as in the original Newton-Leibniz conception. The resulting numbers are called hyperreal numbers, and they can be used to give a Leibniz-like development of the usual rules of calculus.

Significance

While some of the ideas of calculus had been developed earlier in Egypt, Greece, China, India, Iraq, Persia, and Japan, the modern use of calculus began in Europe, during the 17th century, when Isaac Newton and Gottfried Wilhelm Leibniz built on the work of earlier mathematicians to introduce its basic principles.

The development of calculus was built on earlier concepts of instantaneous motion and area underneath curves. Applications of differential calculus include computations involving velocity and acceleration, the slope of a curve, and optimization. Applications of integral calculus include computations involving area, volume, arc length, centre of mass, work, and pressure. More advanced applications include power series and Fourier series. Calculus is also used to gain a more precise understanding of the nature of space, time, and motion.

For centuries, mathematicians and philosophers wrestled with paradoxes involving division by zero or sums of infinitely many numbers. These questions arise in the study of motion and area. The ancient Greek philosopher Zeno of Elea gave several famous examples of such paradoxes. Calculus provides tools, especially the limit and the infinite series, which resolve the paradoxes.

Chapter 3

Linear Differential Equation

In mathematics, a linear differential equation is a differential equation of the form

$$Ly = f$$

where the differential operator L is a linear operator, y is the unknown function, and the right hand side f is a given function (called the source term). The linearity condition on L rules out operations such as taking the square of the derivative of y; but permits, for example, taking the second derivative of y. Therefore a fairly general form of such an equation would be

$$a_n(x)\, D^n\, y(x) + a_{n-1}\,(x)\, D^{n-1}\, y(x)$$
$$+ \ldots + a_1(x)\, Dy(x) + a_0(x)y(x) = f(x)$$

where D is the differential operator d/dx (i.e. $Dy = y'$, $D^2y = y''$,..), and the a_i are given functions. Such an equation is said to have order n, the index of the highest derivative of y that is involved. (Assuming a possibly existing coefficient a_n of this derivative to be non zero, it is eliminated by dividing through it. In case it can become zero, different cases must be considered separately for the analysis of the equation.) If y is assumed to be a function of only one variable, one speaks about an ordinary differential equation, else the derivatives and their coefficients must be understood as (contracted) vectors, matrices or tensors of higher rank, and we have a (linear) partial differential equation. The case where $f = 0$ is called a homogeneous equation and its solutions are called complementary functions. It is particularly important to the solution of the general case, since any complementary function can be added to a solution of the inhomogeneous equation to give another solution (by a method traditionally called *particular integral and complementary function*). When the a_i are numbers, the equation is said to have *constant coefficients.*

The subject of ordinary differential equations encompasses such a large field that you can make a profession of it. There are however a small number of techniques in the subject that you have to know. These are the ones that come up so often in physical systems that you need not only the skills to use them but the intuition about what they will do.

Linear Constant-Coefficient

A differential equation such as

$$\left(\frac{d^2x}{dt^2}\right)^3 + t^2x^4 + 1 = 0$$

is difficult to solve, sinceit is non-linear. This means that if I have two solutions $x_1(t)$ and $x_2(t)$ then the sum $x_1 + x_2$ is not a solution; look at all the cross-terms you get if you try to plug the sum in to the equation. Also if you multiply $x_1(t)$ by 2 you no longer have a solution.

An equation such as

$$e^t \frac{d^3x}{dt^3} + t^2 \frac{dx}{dt} - x = 0$$

may be a mess to solve, but if you have two solutions, $x_1(t)$ and $x_2(t)$ then the sum $\alpha x_1 + \beta x_2$ is also a solution. This is called a linear, homogeneous equation because of this property. A similar-looking equation,

$$e^t \frac{d^3x}{dt^3} + t^2 \frac{dx}{dt} - x = t$$

does not have this property, though it's close. It's called a linear, inhomogeneous equation. If $x_1(t)$ and $x_2(t)$ are solutions to this, then if you try their sum as a solution you get $2t = t$, and that's no solution, but it misses working only because of the single term on the right. One of the most common sorts of differential equations that you see is an especially simple one to solve. That's part of the reason it's so common. This is the linear, constant-coefficient, differential equation. If you have a mass tied to the end of a spring and the other end of the spring is fixed, the force applied to the mass by the spring is to a good approximation proportional to the distance that the mass has moved from its equilibrium position.

If the coordinate x is measured from the mass's equilibrium position, the equation $\vec{F} = m\vec{a}$ says

$$m\frac{d^2x}{dt^2} = -kx$$

If there's friction (and there's *always* friction), the force has another term. Now how do you describe friction mathematically? The common model for dry friction is that the magnitude of the force is independent of the magnitude of the mass's velocity and opposite to the direction of the velocity. If you try to write that down in a compact mathematical form you get something like,

$$\vec{F}_{friction} = -\mu_k F_N \frac{\vec{v}}{|\vec{v}|}$$

This is hard to work with. It can be done, but you are going to do something different. Wet friction is easier to handle mathematically because when you lubricate a surface, the friction becomes velocity dependent in a way that is, for low speeds, proportional to the velocity.

Neither of these two representations is a completely accurate description of the way friction works. That's far more complex than either of these simple models, but these approximations are good enough for many purposes and I'll settle for them. Assume "wet friction" and the differential equation for the motion of m is

$$m\frac{d^2x}{dt^2} = -kx - b\frac{dx}{dt}$$

This is a second order, linear, homogeneous, differential equation, which simply means that the highest derivative present is the second, the sum of two solutions is a solution, and a constant multiple of a solution is a solution.

That the coefficients are constants makes this an easy equation to solve. All you have to do is to recall that the derivative of an exponential is an exponential. $de^t/dt = e^t$. If you substitute this exponential for $x(t)$, of course it can't work as a solution; it doesn't even make sense dimensionally. What is e to the power of a day?

You have to have something in the exponent to make it dimensionless, $e^{\alpha t}$.

Also, the function x is supposed to give you a position, with dimensions of length. Use another constant: $x(t) = Ae^{\alpha t}$. Plug *this* into the differential equation to find

$$mA\alpha^2 e^{\alpha t} + bA\alpha e^{\alpha t} + kAe^{\alpha t} = Ae^{\alpha t}[m\alpha^2 + b\alpha + k] = 0$$

The product of factors is zero, and the only way that a product of two numbers can be zero is if one of the numbers is zero. The exponential never vanishes, and for a non-trivial solution $A \neq 0$, so all that's left is the polynomial in α.

$$m\alpha^2 + b\alpha + k = 0,$$

with solutions $\alpha = \dfrac{-b \pm \sqrt{b^2 - 4km}}{2m}$

The position function is then

$$x(t) = Ae^{\alpha_1 t} + Be^{\alpha_2 t}$$

where A and B are arbitrary constants and α_1 and α_2 are the two roots. Supposed that it is oscillating? It is a harmonic oscillator after all, but the exponentials don't look very oscillatory.

If you have a mass on the end of a spring and the entire system is immersed in honey, it won't do much oscillating! Translated into mathematics, this says that if the constant b is too large, there is no oscillation. In the equation for α, if b is large enough the argument of the square root is positive and both α's are real — no oscillation. Only if b is small enough does the argument of the square root become negative; then you get complex values for the α's and hence oscillations. Push this to the extreme case where the damping vanishes: $b = 0$.

Then $\alpha_1 = i\sqrt{k/m}$ and $\alpha_2 = -i\sqrt{k/m}$.

Denote $\omega_0 = \sqrt{k/m}$.

$$x(t) = Ae^{i\omega_0 t} + Be^{-i\omega_0 t}$$

You can write this in other forms, see problem 10. To determine the arbitrary constant A and B you need two equations. They come from some additional information about the problem, typically some initial conditions. Take a specific example in which you start from the origin with a kick, $x(0) = 0$ and $\dot{x}(0) = v_0$.

$$x(0) = 0 = A + B, \quad \dot{x}(0) = v0 = i\omega_0 A - \omega_0 B$$

Solve for A and B to get $A = -B = v_0/(2i\omega_0)$. Then

$$x(t) = \frac{v_0}{2i\omega_0}[e^{i\omega_0 t} - e^{-i\omega_0 t}] = \frac{v_0}{\omega_0}\sin\omega_0 t$$

As a check on the algebra, use the first term in the power series expansion of the sine function to see how x behaves for small t. The sine factor is sin $\omega_0 t \approx \omega_0 t$, and then $x(t)$ is approximately $v_0 t$, just as it should be. Also notice that despite all the complex numbers, the final answer is real.

Damped Oscillator

If there is damping, but not too much, then the α's have an imaginary part *and* a negative real part.

$$\alpha = \frac{-b \pm i\sqrt{4km - b^2}}{2m} = -\frac{b}{2m} \pm i\omega', \text{ where}$$

$$\omega' = \sqrt{\frac{k}{m} - \frac{b^2}{4m^2}}$$

This represents a damped oscillation and has frequency a bit lower than the one in the undamped case. Use the same initial conditions as above and you will get similar results (let $(\gamma = b/2m)$

$$x(t) = Ae^{(-\gamma+i\omega')t} + Be^{(-\gamma-i\omega')t}$$

$$x(0) = A + B = 0,\ v_x(0) = (-\gamma + i\omega')A + (-\gamma - i\omega')B = v_0$$

For small values of t, the first terms in the power series expansion of this result are

$$x(t) = \frac{v_0}{\omega'}[1 - \gamma t + \gamma^2 t^2 / 2 - ...]$$

$$[\omega' t - \omega'^3 t^3 / 6 + ...] = v_0 t - v_0 \gamma t^2 + ...$$

The first term is what you should expect, as the initial velocity is $v_x = v_0$. The negative sign in the next term says that it doesn't move as far as it would without the damping, but analyze it further. Does it have the right size as well as the right sign? It is $-v_0\gamma t^2 = -v_0(b/2m)t^2$. But that's an acceleration: $a_x t^2/2$. It says that the acceleration just after the motion starts is $a_x = -bv_0/m$. Is that what you should expect? As the motion starts, the mass hasn't gone very far so the spring doesn't yet exert much force. The viscous friction is however $-bv_x$. Set that equal to ma_x and you see that,

$$-v_0\gamma t^2$$

has precisely the right value. In Equation you assumed that the two roots of the quadratic, the two α's, are different. What if they aren't? Then you have only one value of to use in defining the solution $e^{\alpha t}$ in Equation. You now have only one arbitrary constant with which to match two initial conditions. What is the energy for this damped oscillator? The kinetic energy is $mv^2/2$ and the potential energy for the spring is $kx^2/2$. Is the sum constant? No.

$$\frac{d}{dt}\frac{1}{2}(mv^2 + kx^2) = mv\frac{dv}{dt} + kx\frac{dx}{dt}$$

$$= v_x(ma_x + kx) = -bv_x^2 = F_{x,\text{frict}}\, v_x$$

"Force times velocity" is a common expression for power, and this says that the total energy is decreasing according to this formula. The energy decreases exponentially on average.

Sequences of Functions

Let's start off this section with a discussion of just what a sequence is. A sequence is nothing more than a list of numbers written in a specific order. The list may or may not have an infinite number of terms in them although we will be dealing exclusively with infinite sequences in this class. General sequence terms are denoted as follows,

$$\begin{aligned} &a_1 - \text{first term} \\ &a_2 - \text{second term} \\ &\vdots \\ &a_n - n^{th} \text{ term} \\ &a_{n+1} - (n+1)^{st} \text{ term} \\ &\vdots \end{aligned}$$

Because we will be dealing with infinite sequences each term in the sequence will be followed by another term as noted above. In the notation above we need to be very careful with the subscripts. The subscript of $n + 1$ denotes the next term in the sequence and NOT one plus the nth term! In other words,

$$a_{n+1} \neq a_n + 1$$

So be very careful when writing subscripts to make sure that the "+1" doesn't migrate out of the subscript! This is an easy mistake to make when you first start dealing with this kind of thing.

There is a variety of ways of denoting a sequence. Each of the following are equivalent ways of denoting a sequence.

$$\{a_1, a_2, \ldots, a_n, a_{n+1}, \ldots\} \qquad \{a_n\} \qquad \{a_n\}_{n+1}^{\infty}$$

In the second and third notations above a_n is usually given by a formula. A couple of notes are now in order about these notations. First, note the difference between the second and third notations above. If the starting point is not important or is implied in some way by the problem it is often not written down as we did in the third notation.

Next, we used a starting point of $n = 1$ in the third notation only so we could write one down. There is absolutely no reason to believe that a sequence will start at $n = 1$. A sequence will start where ever it needs to start. Let's take a look at a couple of sequences.

Example: Write down the first few terms of each of the following sequences.

1. $\left\{\dfrac{n+1}{n^2}\right\}_{n=1}^{\infty}$ 2. $\left\{\dfrac{(-1)^{n+1}}{2^n}\right\}_{n=0}^{\infty}$

3. $\{b_n\}_{n=0}^{\infty}$, where $b_n = n^{th}$ digit of π

Solution:

1. $\left\{\dfrac{n+1}{n^2}\right\}_{n=1}^{\infty}$

To get the first few sequence terms here all we need to do is plug in values of n into the formula given and we'll get the sequence terms.

$$\left\{\frac{n+1}{n^2}\right\}_{n=1}^{\infty} = \left\{\underbrace{2}_{n=1}, \underbrace{\frac{3}{4}}_{n=2}, \underbrace{\frac{4}{9}}_{n=3}, \underbrace{\frac{5}{16}}_{n=4}, \underbrace{\frac{6}{25}}_{n=5}, \dots\right\}$$

Note the inclusion of the "..." at the end! This is an important piece of notation as it is the only thing that tells us that the sequence continues on and doesn't terminate at the last term.

2. $\left\{\dfrac{(-1)^{n+1}}{2^n}\right\}_{n=0}^{\infty}$

This one is similar to the first one. The main difference is that this sequence doesn't start at $n = 1$.

$$\left\{\frac{(-1)^{n+1}}{2^n}\right\}_{n=0}^{\infty} = \left\{-1, \frac{1}{2}, \frac{1}{4}, \frac{1}{8}, \frac{1}{16}, \dots\right\}$$

Note that the terms in this sequence alternate in signs. Sequences of this kind are sometimes called alternating sequences.

3. $\{b_n\}_{n=1}^{\infty}$, where $b_n = n^{th}$ digit of π

This sequence is different from the first two in the sense that it doesn't have a specific formula for each term. However, it does tell us what each term should be. Each term should be the nth digit of π. So we know that p = 3.14159265359...

The sequence is then,

$$\{3, 1, 4, 1, 5, 9, 2, 6, 5, 3, 5,...\}$$

In the first two parts of the previous example note that we were really treating the formulas as functions that can only have integers plugged into them. Or,

$$f(n) = \frac{n+1}{n^2} \qquad g(n) = \frac{(-1)^{n+1}}{2^n}$$

This is an important idea in the study of sequences (and series). Treating the sequence terms as function evaluations will allow us to do many things with sequences that couldn't do otherwise.

Before delving further into this idea however we need to get a couple more ideas out of the way.

First we want to think about "graphing" a sequence. To graph the sequence $\{a_n\}$ we plot the points $(n_3\ a_n)$ as n ranges over all possible values on a graph.

For instance, let's graph the sequence $\left\{\frac{n+1}{n^2}\right\}_{n=1}^{\infty}$. The first few points on the graph are,

$$(1, 2), \left(2, \frac{3}{3}\right), \left(3, \frac{4}{9}\right), \left(4, \frac{5}{16}\right), \left(5, \frac{6}{25}\right), ...$$

The graph, for the first 30 terms of the sequence, is then,

This graph leads us to an important idea about sequences. Notice that as n increases the sequence terms from our sequence terms, in this case, get closer and closer to zero. We then say that zero is the limit (or sometimes the limiting value) of the sequence and write,

$$\lim_{n\to\infty} a_n = \lim_{n\to\infty} \frac{n+1}{n^2} = 0$$

This notation should look familiar to you. It is the same notation we used when we talked about the limit of a function. In fact, if you recall, we said earlier that we could think of sequences as functions in some way and so this notation shouldn't be too surprising. Using the ideas that we developed for limits of functions we can write down the following *working definition* for limits of sequences.

Working Definition of Limit

1. We say that

   $$\lim_{n\to\infty} a_n = L$$

   if we can make an as close to L as we want for all sufficiently large n. In other words, the value of the a_n's approach L as n approaches infinity.

2. We say that

   $$\lim_{n\to\infty} a_n = \infty$$

   if we can make a_n as large as we want for all sufficiently large n. Again, in other words, the value of the a_n's get larger and larger without bound as n approaches infinity.

3. We say that

   $$\lim_{n\to\infty} a_n = -\infty$$

If we can make an as large and negative as we want for all sufficiently large n.

Again, in other words, the value of the a_n's are negative and get larger and larger without bound as n approaches infinity. The working definitions of the various sequence limits are nice in that they help us to visualize what the limit actually is. Just like with limits of functions however, there is also a precise definition for each of these limits. Let's give those before proceeding.

Precise Definition of Limit

1. We say that $\lim_{n\to\infty} a_n = L$ if for every number Ý > 0 there is an integer N such that

$|an - L| < ⥣$ whenever $n > N$

2. We say that $\lim_{n\to\infty} a_n = \infty$ if for every number $M > 0$ there is an integer N such that
$an > M$ whenever $n > N$
3. We say that $\lim_{n\to\infty} a_n$ if for every number $M < 0$ there is an integer N such that
$an < M$ whenever $n > N$

We won't be using the precise definition often, but it will show up occasionally.

Note that both definitions tell us that in order for a limit to exist and have a finite value all the sequence terms must be getting closer and closer to that finite value as n increases. Now that we have the definitions of the limit of sequences out of the way we have a bit of terminology that we need to look at.

If $\lim_{n\to\infty} a_n$ exists and is finite we say that the sequence is convergent. If $\lim_{n\to\infty} a_n$ doesn't exist or is infinite we say the sequence diverges. Note that sometimes we will say the sequence diverges to $-\infty$ if $\lim_{n\to\infty} a_n = \infty$ and if $\lim_{n\to\infty} a_n$ we will sometimes say that the sequence diverges to $-\infty$.

Get used to the terms "convergent" and "divergent" as we'll be seeing them quite a bit throughout this chapter.

So just how do we find the limits of sequences? Most limits of most sequences can be found using one of the following theorems.

Theorem: Given the sequence $\{a_n\}$ if we have a function $f(x)$ such that $f(n) = a_n$ an $\lim_{n\to\infty} f(x) = L$ then $\lim_{n\to\infty} a_n = L$

This theorem is basically telling us that we take the limits of sequences much like we take the limit of functions. In fact, in most cases we'll not even really use this theorem by explicitly writing down a function. We will more often just treat the limit as if it were a limit of a function and take the limit as we always did back in Calculus I when we were taking the limits of functions. So, now that we know that taking the limit of a sequence is nearly identical to taking the limit of a function we also know that all the properties from the limits of functions will also hold.

Properties

1. $\lim\limits_{n\to\infty} (a_n \pm b_n) = \lim\limits_{n\to\infty} a_n \pm \lim\limits_{n\to\infty} b_n$

2. $\lim\limits_{n\to\infty} ca_n = c \lim\limits_{n\to\infty} a_n$

3. $\lim\limits_{n\to\infty} (a_n b_n) = \left(\lim\limits_{n\to\infty} a_n\right)\left(\lim\limits_{n\to\infty} b_n\right)$

4. $\lim\limits_{n\to\infty} \dfrac{a_n}{b_n} = \dfrac{\lim\limits_{n\to\infty} a_n}{\lim\limits_{n\to\infty} b_n}$, provided $\lim\limits_{n\to\infty} b_n \neq 0$

5. $\lim\limits_{n\to\infty} a_n^p = \left[\lim\limits_{n\to\infty} a_n\right]^p$ provided $a_n \geq 0$ provided

These properties can be proved using Theorem above and the function limit properties we saw in Calculus I or we can prove them directly using the precise definition of a limit using nearly identical proofs of the function limit properties. Next, just as we had a Squeeze Theorem for function limits we also have one for sequences and it is pretty much identical to the function limit version.

Squeeze Theorem for Sequences

If $a_n \leq c_n \leq b_n$ for all $n > N$ for some N and $\lim\limits_{n\to\infty} a_a$ $\lim\limits_{n\to\infty}$ $b_n = L$ then $\lim\limits_{n\to\infty} c_n = L$.

Note that in this theorem the "for all $n > N$ for some N" is really just telling us that we need to have $a_n \leq c_n \leq b_n$ for all sufficiently large n, but if it isn't true for the first few n that won't invalidate the theorem.

As we'll see not all sequences can be written as functions that we can actually take the limit of. This will be especially true for sequences that alternate in signs. While we can always write these sequence terms as a function we simply don't know how to take the limit of a function like that. The following theorem will help with some of these sequences.

Theorem: If $\lim_{n\to\infty} |a_n| = 0$ then $\lim_{n\to\infty} a_n = 0$.

Note that in order for this theorem to hold the limit MUST be zero and it won't work for a sequence whose limit is not zero. This theorem is easy enough to prove so let's do that. Proof of Theorem: The main thing to this proof is to note that,

$$-|a_n| \le a_n \le |a_n|$$

Then note that,

$$\lim_{n\to\infty} (-|a_n|) = -\lim_{n\to\infty} |a_n| = 0$$

We then have $\lim_{n\to\infty} (-|a_n|) = \lim_{n\to\infty} |a_n| = 0$ and so by the Squeeze Theorem we must also have,

$$\lim_{n\to\infty} a_n = 0$$

The next theorem is a useful theorem giving the convergence/divergence and value (for when it's convergent) of a sequence that arises on occasion.

Theorem: The sequence $\left\{r^n\right\}_{n=0}^{\infty}$ converges if $-1 < r \le 1$ and diverges for all other value of r. Also,

$$\lim_{n\to\infty} r^n = \begin{cases} 0 & \text{if } -1 < r < 1 \\ 1 & \text{if } r = 1 \end{cases}$$

Partial Proof of Theorem: We'll do this by a series of cases although the last case will not be completely proven.

Case 1: r > 1

We know from Calculus I that $\lim_{n\to\infty} r^x = \infty$ if $r > 1$ and so by Theorem above we also know that $\lim_{n\to\infty} r^n = \infty$ and so the sequence diverges if $r > 1$.

Case 2: r = 1

In this case we have,

$$\lim_{n\to\infty} r^n = \lim_{n\to\infty} 1^n = \lim_{n\to\infty} 1 = 1$$

So, the sequence converges for $r = 1$ and in this case its limit is 1.

Case 3: 0 < r < 1

We know from Calculus I that $\lim_{n\to\infty} r^x = 0$ if $0 < r < 1$ and so by Theorem above we also know that $\lim_{n\to\infty} r^n = 0$ and so the sequence converges if $0 < r < 1$ and in this case its limit is zero.

Case 4: r = 0

In this case we have,

$$\lim_{n\to\infty} r^n = \lim_{n\to\infty} 0^n = \lim_{n\to\infty} 0 = 0$$

So, the sequence converges for $r = 0$ and in this case its limit is zero.

Case 5: – 1 < r < 0

First let's note that if $-1 < r < 0$ then $0 < |r| < 1$ then by Case 3 above we have,

$$\lim_{n\to\infty} |r^n| = \lim_{n\to\infty} |r^n| = 0$$

Theorem above now tells us that we must also have, $\lim_{n\to\infty} r^n = 0$ and so if $-1 < r < 0$ the sequence converges and has a limit of 0.

Case 6:

In this case the sequence is,

$$\left\{r^n\right\}_{n=0}^{\infty} = \left\{(-1)^n\right\}_{n=0}^{\infty}$$

$$= \{1,-1,1,-1,1-1,1-1,\ldots\}_{n=0}^{\infty}$$

and hopefully it is clear that $\lim_{n \to \infty} (-1)^n$ doesn't exist. Recall that in order of this limit to exist the terms must be approaching a single value as n increases. In this case however the terms just alternate between 1 and –1 and so the limit does not exist.

So, the sequence diverges for $r = -1$.

Case 7: r < – 1

In this case we're not going to go through a complete proof. Let's just see what happens if we let $r = -2$ for instance. If we do that the sequence becomes,

$$\left\{r^n\right\}_{n=0}^{\infty} = \left\{(-2)^n\right\}_{n=0}^{\infty} = \{1, -2, 4, -8, 16, -32, ...\}_{n=0}^{\infty}$$

So, if $r = -2$ we get a sequence of terms whose values alternate in sign and get larger and larger and so $\lim_{n \to \infty} (-2)^n$ doesn't exist. It does not settle down to a single value as n increases nor do the terms ALL approach infinity. So, the sequence diverges for $r = -2$. We could do something similar for any value of r such that $r < -1$ and so the sequence diverges for $r < -1$.

Let's take a look at a couple of examples of limits of sequences.

Example: Determine if the following sequences converge or diverge. If the sequence converges determine its limit.

1. $\left\{\dfrac{3n^2 - 1}{10n + 5n^2}\right\}_{n=2}^{\infty}$
2. $\left\{\dfrac{e^{2n}}{n}\right\}_{n=1}^{\infty}$
3. $\left\{\dfrac{(-1)^n}{n}\right\}_{n=1}^{\infty}$
4. $\left\{(-1)^n\right\}_{n=1}^{\infty}$

Solution:

1. $\left\{\dfrac{3n^2 - 1}{10n + 5n^2}\right\}_{n=2}^{\infty}$

In this case all we need to do is recall the method that was developed in Calculus I to deal with the limits of rational functions

To do a limit in this form all we need to do is factor from the numerator and denominator the largest power of n, cancel and then take the limit.

$$\lim_{n\to\infty}\frac{3n^2-1}{10n+5n^2}=\lim_{n\to\infty}\frac{n^2\left(3-\frac{1}{n^2}\right)}{n^2\left(\frac{10}{n}+5\right)}=\lim_{n\to\infty}\frac{3-\frac{1}{n^2}}{\frac{10}{n}+5}=\frac{3}{5}$$

So the sequence converges and its limit is $\frac{3}{5}$.

2. $\left\{\frac{e^{2n}}{n}\right\}_{n=1}^{\infty}$

We will need to be careful with this one. We will need to use L'Hospital's Rule on this sequence. The problem is that L'Hospital's Rule only works on functions and not on sequences. Normally this would be a problem, but we've got Theorem from above to help us out. Let's define

$$f(x)=\frac{e^{2x}}{x}$$

and note that,

$$f(n)=\frac{e^{2n}}{n}$$

Theorem says that all we need to do is take the limit of the function.

$$\lim_{n\to\infty}\frac{e^{2n}}{n}=\lim_{n\to\infty}\frac{e^{2x}}{x}=\lim_{n\to\infty}\frac{2e^{2x}}{1}=\infty$$

So, the sequence in this part diverges (to ∞).

More often than not we just do L'Hospital's Rule on the sequence terms without first converting to x's since the work will be identical regardless of whether we use x or n. However, we really should remember that technically we can't do the derivatives while dealing with sequence terms.

3. $\left\{\frac{(-1)^n}{n}\right\}_{n=1}^{\infty}$

We will also need to be careful with this sequence. We might be tempted to just say that the limit of the sequence terms is zero (and we'd be correct). However, technically we can't take the limit of sequences whose terms alternate in sign, because we don't know how to do limits of functions that exhibit that same behavior.

Also, we want to be very careful to not rely too much on intuition with these problems. So, let's work this one by the book. We will need to use Theorem on this problem. To this we'll first need to compute,

$$\lim_{n\to\infty}\left|\frac{(-1)^n}{n}\right| = \lim_{n\to\infty}\frac{1}{n} = 0$$

Therefore, since the limit of the sequence terms with absolute value bars on them goes to zero we know by Theorem that,

$$\lim_{n\to\infty}\frac{(-1)^n}{n} = 0$$

Which also means that the sequence converges to a value of zero.

4. $\left\{(-1)^n\right\}_{n=0}^{\infty}$

For this theorem note that all we need to do is realize that this is the sequence in Theorem above using $r = -1$. So, by Theorem this sequence diverges. We now need to give a warning about misusing Theorem. Theorem only works if the limit is zero. If the limit of the absolute value of the sequence terms is not zero then the theorem will not hold. The last part of the previous example is a good example of this (and in fact this warning the whole reason that part is there). Notice that

$$\lim_{n\to\infty}\left|(-1)^n\right| = \lim_{n\to\infty} 1 = 1$$

and yet, $\lim_{n\to\infty} (-1)^n$ doesn't even exist let alone equal 1.So, be careful using this Theorem. Before moving onto the next section we need to give one more theorem that we'll need for a proof down the road.

Theorem: For the sequence $\{a_n\}$ if both $\lim_{n\to\infty} a_{2n} = L$ and $\lim_{n\to\infty} a_{2n+1} = L$ then $\{a_n\}$ is convergent and $\lim_{n\to\infty} a_n = L$.

Proof of Theorem

Let $\Sigma > 0$.

Then since $\lim_{n\to\infty} a_{2n} = L$ there is an $N_1 > 0$ such that if $n > N_1$ we know that,

$$|a2n - L| < å$$

Likewise, because,

$$\lim_{n\to\infty} a_{2n+1} = L$$

there is an $N_2 > 0$ such that if $n > N_2$ we know that,

$$|a_{2n+1} - L| < å$$

Now, let $N = \max\{2N_1, 2N_2 + 1\}$ and let $n > N$. Then either $an = a_{2k}$ for some $k > N_1$ or $a_n = a_{2k+1}$ for some $k > N_2$ and so in either case we have that,

$$|a_n - L| < å$$

Therefore, $\lim_{n\to\infty} a_n = L$ and so $\{a_n\}$ is convergent.

More on Sequences

In the previous section we introduced the concept of a sequence and talked about limits of sequences and the idea of convergence and divergence for a sequence.

In this section we want to take a quick look at some ideas involving sequences.

Let's start off with some terminology and definitions.

Given any sequence $\{a_n\}$ we have the following.

1. We call the sequence increasing if $a_n < a_{n+1}$ for every n.
2. We call the sequence decreasing if $a_n > a_{n+1}$ for every n.
3. If $\{a_n\}$ is an increasing sequence or $\{a_n\}$ is a decreasing sequence we call it monotonic.
4. If there exists a number m such that m £ a_n for every n we say the sequence is bounded below. The number m is sometimes called a lower bound for the sequence.
5. If there exists a number M such that a_n £ M for every n we say the sequence is bounded above. The number M is sometimes called an upper bound for the sequence.
6. If the sequence is both bounded below and bounded above we call the sequence bounded.

Note that in order for a sequence to be increasing or decreasing it must be increasing/decreasing for every n. In other words, a sequence that increases for three terms and then decreases for the rest of the terms is NOT a decreasing sequence! Also note that a monotonic sequence must always increase or it must always decrease. Before moving on we should make a quick point about the bounds for a sequence that is bounded above and/or below. We'll make the point about lower bounds, but we could just as easily make it about upper bounds. A sequence is bounded below if we can find any number m such that $m \leq a_n$ for every n. Note however that if we find one number m to use for a lower bound then any number smaller than m will also be a lower bound. Also, just because we find one lower bound that doesn't mean there won't be a "better" lower bound for the sequence than the one we found. In other words, there are an infinite number of lower bounds for a sequence that is bounded below, some will be better than others. In my class all that I'm after will be a lower bound. I don't necessarily need the best lower bound, just a number that will be a lower bound for the sequence.

Chapter 4

Theories of Differentiation

Multiplication by a Constant

If c is a constant and $f(x) = cf(x)$, then $f(x)$ is differentiableand

$$\phi'(x) = cf'(x)$$

This follows immediately from the relation

$$\frac{\phi(x+h)-\phi(x)}{h} = c\frac{f(x+h)-f(x)}{h}$$

if we take the limit $h \rightarrow 0$.

Differentiation of the Rational Functions

To begin with, we shall again deduce the differentiation formula,

$$\frac{d}{dx}x^n = nx^{n-1}$$

for every positive integer n, basing the proof on the rule for differentiation of a product. x and hence obtain

$$\frac{d}{dx}x^n = 1.x^{n-1} + 1.x^{n-1} + \ldots + 1.x^{n-1}$$

$$= nx^{n-1}.$$

Continuing this process, we obtain

$$\frac{d^3}{dx^3}x^n = n(n-1)(n-2)x^{n-3}$$

$$\frac{d^3}{dx^3}x^n = 1.2...n = n!$$

It is clear from the last of these formulae that the $(n + 1)$-th derivative of x^n vanishes everywhere. By virtue of our first two rules, a knowledge of the differentiation of powers enables us at once to differentiate any polynomial

$$y = a_0 + a_1x + a_2x + ... + a_nx^n.$$

We have simply

$$y' = a_1 + 2a_2x + 3a_3x^2 + ... + na_nx^{n-1}$$

and

$$y'' = 2a_2 + 3.2a_3x + 4.3a_4x^2 + ... + n(n-1)a_nx^{n-2}$$

and so on.

The differentiation of any rational function now follows with the help of thequotient rule. In particular, we shall again derive the differentiation formula for the function x^n, where $n = -m$ is a negative integer. The application of thequotient rule together with the fact that the derivative of a constant is equal to zero yields

$$\frac{d}{dx}\left(\frac{1}{x^m}\right) = -\frac{mx^{m-1}}{x^{2m}} = -\frac{m}{x^{m+1}}$$

or, if we take $m = -n$,

$$\frac{d}{dx}x^n = nx^{n-1}$$

which agrees formally with the result for positive values of n and with the results given earlier.

The Derivatives of a Function of Several Variables

Definition. Geometrical Representation

If we assign in a function of several variables definite numerical values to all but one of the variables and allow only that one variable, say x, to vary, the function becomes a function of one variable.

The function $u = f(x, y_0)$ of the single variable x, which is thus formed, may be simply represented geometrically by letting the surface $u = f(x, y)$ be cut by the plane $y = y_0$. The curve of intersection thus formed in the plane is represented by the equation $u = f(x,y_0)$.

If we differentiate this funtion in the usual way at the point $x = x_0$ (assuming that the derivative exists), we obtain the partial derivative of $f(x,y)$ with respect to x at the point (x_0, y_0).

According to the ordinary definition of the derivative, this is the limit,

$$\lim_{h\to 0} \frac{f(x_0+h, y_0) - f(x_0, y_0)}{h}$$

If (x_0, y_0) is a point on the boundary of the region of definition, we make the restriction that in the passage to the limit the point,

$$(x + h, y_0)$$

must always remain in the region. Geometrically speaking, this partial derivative denotes the tangent of the angle between a parallel to the x-axis and the tangent line to the curve $u = (x,y_0)$.

It is therefore the slope of the surface u = $f(x,y)$ in the direction of the x-axis. In order to represent these partial derivatives, several different notations are in use of which we mention:

$$\lim_{h\to 0} \frac{f(x_0+h, y_0) - f(x_0, y_0)}{h}$$

$$= f_\infty(x_0, y_0) = u_\infty(x_0, y_0)$$

If we wish to emphasize -that the partial derivative is the limit of a difference quotient, we denote it by

$$\frac{\partial f}{\partial x} or \frac{\partial}{\partial x} f \, .$$

We use here a special curved letter ∂ instead of the ordinary *d* used in the differentiation of functions of one variable, in order to show that we are dealing with a function of several variables and differentiating with respect to one of them. Sometimes. it is convenient to use Cauchy's symbol *D* and to write,

$$\frac{\partial f}{\partial x} = D_{\infty} f;$$

however, we shall rarely use this symbol.

In exactly the same way, we define the partial derivative
o f
$f(x, y)$ with respect to y at the point (x_0, y_0) by the relation

$$\lim_{h \to 0} \frac{f(x_0, y_0 + k) - f(x_0, y_0)}{k} = f_v(x_0, y_0)$$

$$= \frac{\partial f}{\partial y} = D_v f(x_0, y_0)$$

It represents the slope of the curve of intersection of the surface $u = f(x, y)$ with the plane $x = x_0$, perpendicular to the x-axis. Now, let us think of the point (x_0, y_0), hitherto considered to be fixed, as variable and accordingly omit the suffices. In other word, we think of the differentiation as being carried out at any point (x, y) of the region of definition of $f(x, y)$. Then the two derivatives are themselves functions of x and y:

$$u_x(x, y) = f_x(x, y) = \frac{\partial f(x, y)}{\partial x} \text{ and}$$

$$u_y(x, y) = f_y(x, y) = \frac{\partial f(x, y)}{\partial y}$$

We likewise denote the third partial derivatives by

$$\frac{\partial}{\partial x}\left(\frac{\partial^2 f}{\partial x^2}\right) = \frac{\partial^2 f}{\partial x^3} = f_{xxx},$$

$$\frac{\partial}{\partial y}\left(\frac{\partial^2 f}{\partial x^2}\right) = \frac{\partial^3 f}{\partial y dx^2} = f_{yxx},$$

$$\frac{\partial}{\partial x}\left(\frac{\partial^2 f}{\partial x \partial y}\right) = \frac{\partial^2 f}{\partial x^2 \partial y} = f_{xxy}, \text{ etc.}$$

and, in general, the n-th derivatives

$$\frac{\partial}{\partial x}\left(\frac{\partial^{n-1} f}{\partial x^{n-1}}\right) = \frac{\partial^n f}{\partial x^n} = fx^n$$

$$\frac{\partial}{\partial y}\left(\frac{\partial^{n-1} f}{\partial x^{n-1}}\right) = \frac{\partial^n f}{\partial y \partial x^{n-1}} = fyx^{n-1}, \&c.$$

Finally, we shall study a few examples of the actual calculation of partial derivatives.

According to the definition, all the independent variables are to be kept constant except the one with respect to which we are differentiating. We therefore have merely to regard the other variables as constants and carry out out the differentiation according to the rules by which we differentiate functions of a single independent variable.

For example, we have:

1. Function $f(x, y) = xy$

   First Derivatives, $f_x = y, f_v = x$

   Second Derivatives, $f_{xx} = 0, f_{xy} = f_{yx} = 1, f_{yy} = 0$

2. Function $f(x, y) = \sqrt{(x^2 + y^2)}$

1. Reciprocal of the radius vector in three dimensions:

$$f(x, y, z) = \frac{1}{\sqrt{(x^2 + y^2 + z^2)}} = \frac{1}{r}$$

First Derivatives,

$$f_x = -\frac{x}{\sqrt{(x^2 + y^2 + z^2)^2}} = -\frac{x}{r^2}$$

$$f_y = -\frac{x}{\sqrt{\left(x^2+y^2+z^2\right)^2}} = -\frac{y}{r^2}$$

$$f_z = -\frac{x}{\sqrt{\left(x^2+y^2+z^2\right)^2}} = -\frac{z}{r^2}$$

Second Derivatives,

$$f_{xx} = -\frac{1}{r^2}+\frac{3x^2}{r^2}, f_{yy} = -\frac{1}{r^2}+\frac{3y^2}{r^2}, f_{zz} = -\frac{1}{r^2}+\frac{3z^2}{r^2}$$

$$f_{xy} = f_{yx} = \frac{3xy}{r^2}, f_{yz} = f_{zy} = \frac{3yz}{r^2}, f_{zx} = f_{zx} = \frac{3zx}{r^2}$$

We see from this that there holds for the function

$$f = \frac{1}{\sqrt{\left(x^2+y^2+z^2\right)}}$$ the equation,

$$f_{xx}+f_{yy}+f_{zz} = -\frac{3}{r^2}+\frac{3\left(x^2+y^2+z^2\right)}{r^2} = 0,$$

for all values of x, y, z except 0,0,0; as we may say, the equation

$$f_{xx}+f_{yy}+f_{zz} = 0$$

is satisfied identically in x, y, z by the function $f(x, y, z) = 1/r$.

2. Functions, $f(x,y) = \frac{1}{\sqrt{y}} e^{-(x-a)^2/4y}$

First Derivatives,

$$f_x = \frac{1}{\sqrt{y}} \frac{-(x-a)}{2y} e^{-(x-a)^2/4y}$$

$$f_y = \left(\frac{-1}{2y^{3/2}}+\frac{(x-a)^2}{4y^{3/2}}\right) e^{-(x-a)^2/4y}$$

Second Derivatives,

$$f_{xx} = \left(\frac{-1}{2y^{3/2}} + \frac{(x-a)^2}{4y^{3/2}} \right) e^{-(x-a)^2/4y}$$

The equation

$$f_{xx} - f_y = 0$$

is therefore satisfied identically in x and y.

Just as in the case of one independent variable, the possession of derivatives is a special property of a function. Note that the expression differentiable implies more than that the partial derivatives with respect to x and y exist. All the same, this property is possessed by all functions of practical importance except perhaps at isolated exceptional points. In contrast to functions of one variable, the possession of derivatives does not imply the continuity of the function. This is clearly shown by the example $u = \dfrac{2xy}{x^2 + u^2}$ already considered. because the partial derivatives exist for it everywhere and yet the function is discontinuous at the origin. But, as is stated by the following theorem, the possession of bounded derivatives does imply continuity: If a function $f(x,y)$ has partial derivatives f_x and f_y everywhere in a region R, and these derivatives satisfy everywhere the in equalities

$$|f_x(x,y)| < M, |f_x(x,y)| < M,$$

where M is independent of x and y, then $f(x, y)$ is continuous everywhere in R. In particular, if f_x and f_y are continuous, they are necessarily bounded, so that $f(x,y)$ is also continuous. In other words, it made no difference whether we differentiated first with respect to x and then with respect to y or *vice versa*. This is no accidental occurrence.

Theorem: If the mixed partial derivatives f_{xy} and f_{yx} of a function $f(x,y)$ are continuous in a region R, then the equation

$$f_{yx} = f_{xy}$$

holds everywhere in the interior of this region, i.e., the order of differentiation with respect to x and y is immaterial.

Differentiation of the Trigonometric Functions

For the trigonometric functions sin *x* and cos *x*, we have already obtained the differentiation formulae

$$\frac{d}{dx}\sin x = \cos x$$

and

$$\frac{d}{dx}\cos x = -\sin x$$

The quotient rule now enables us to differentiate the functions,

$$y = \tan x = \frac{\sin x}{\cos x} \text{ and}$$

$$y = \cot x = \frac{\cos x}{\sin x}$$

According to the rule, the derivative of the first of these functions is

$$y' = \frac{\cos^2 x + \sin^2 x}{\cos^2 x}$$

$$= \frac{1}{\cos^2 x'}$$

and we obtain the result

$$\frac{d}{dx}\tan x = \frac{1}{\cos^2 x}$$

$$= \sec^2 x$$

$$= 1 + \tan^2 x$$

Similarly, we obtain

$$\frac{d}{dx}\cot x = -\frac{1}{\sin^2 x}$$

$$= -\cos ec^2 x$$

Differentiation of Inverse Functions

Functions of Functions (Compound Functions)

It often happens that a function u of the independent variables x, y is stated in the form,

$$u = f(\xi, \eta, \ldots),$$

where the arguments ξ, η,... of the function f are themselves functions of x and y:

$$\xi = \phi(x, y),\ \eta = \psi(x, y), \ldots$$

We then say that,

$$u = f(\mathrm{x}, \eta, \ldots),\ f = (f(x, y), \ldots y(x, y), \ldots) = F(x, y)$$

is given as a compound function of x and y.

For example, the function

$$u = e^{x^3} y(x+y)^3$$

may be written as a compound function by means of the relations

$$u = e^{\xi}\eta^3 = f(\xi, \eta);\ \xi = x^2 y,\ \eta = x + y \cdot$$

Similarly, the function

$$u = \log(x + 1).\ \text{arc cos}\ \sqrt{\left(4 - x^2 - y^2\right)}$$

can be expressed in the form

$$u = \eta \operatorname{arc} \cos \xi = f(\xi, \eta),$$

$$\xi = \sqrt{\left(4 - x^2 - y^2\right)}, \eta = \log(x+1).$$

In order to make this concept more precise, we assume to begin with that the functions $\xi = \phi(x, y)$, $\eta = \psi(x, y)$,... are defined in a certain region R of the independent variables x, y. Then there corresponds to every point (x,y) of R a point $(\xi, \eta, ..)$. As the point (x,y) ranges over R, the point $(\xi, \eta, \ldots)$ will range over a certain set of values. We assume that the point $(\xi, \eta, \ldots)$ always lies within a region S in which $f(\xi, \eta, \ldots)$ is defined. The function

$$u = f(x, y), \psi(x, y),...) = F(x, y)$$

is then defined in the region R.

Referring to our examples, we find in the first one of them that ξ and η are defined for every x, y and $f(\xi, \eta)$ is defined for every ξ, η, so that our region R can be taken to be the entire xy-plane. However, in the second example, the region S is restricted by the inequality $|\xi| \leq 1$, since for $|\xi| > 1$ the function across ξ is undefined.

Secondly, the region R is restricted by the inequalities $x+1>0$ and $x + y \leq 4$, since for other values both ξ and η are not defined. Thirdly, the region R must be further limited by the inequality $3 \leq x + y$ in order that the point with co-ordinates ξ, η shall fall into S, i.e., the restriction $|\xi| \leq 1$ implies that $x + y \leq 3$. Hence R consists of the part of the ring $3 \leq x+y \leq 4$ lying to the right of the line $x = -1$.

The following theorem on compound functions is an immediate consequence of the definitions: If the function $u = f(\xi, \eta, \ldots)$ is continuous in S and the functions $\xi = \phi(x, y)$, $\eta = \psi(x, y))$, are continuous in R, then the compound function $u = F(x, y)$ is continuous in $R.\xi = \phi(x, y)$. The reader should be able to prove this statement

The Chain Rule

We now turn our attention to compound functions of the type $u = f(\xi, \eta, \ldots)$, where $\xi, \eta, \ldots$ depend on the single variable x:

$$\xi = \phi(x), \eta = \psi(x),....$$

For such functions, we have the important theorem known as the chain rule: If the function $u = f(\xi, \eta, \ldots)$ has continuous partial derivatives of the first order in S and the functions $\xi = \phi(x)$, $\eta = \psi(x), \ldots$ have continuous first derivatives in the interval R, $a \leq x \leq b$, then $u = f(\phi(x), \psi(x), \ldots) = F$ xi) has a continuous derivative in R and

$$F'(x) = f_{\xi}\phi'(x) + f_{\eta}\eta'(x) +....$$

The right-hand side of this equation is an abbreviation for

$$f_x \{f(x), y(x),...\} f'(x) +...$$

In order to simplify the notation, we shall assume that f is a function of the three arguments ξ, η, ζ.

We shall denote by x_0 an arbitrary fixed point of the interval $a \nmid x \nmid b$, by ξ_0, η_0, ζ_0 the corresponding values $\xi_0 = \phi(x_0)$, $\eta_0 = \psi(x_0)$, $\xi_0 = \chi(x_0)$ and by ξ, η, ζ the values $\phi(x)$, $\psi(x)$, $\chi(x)$, corresponding to a variable point $x = x_0 + h$. We first write down the identity,

$$F(x) - F(x_0) = f(\xi, \eta, \zeta) - f(\xi_0, \eta_0, \zeta_0)$$

$$= f(\xi, \eta, \zeta) - f(\xi_0, \eta_0, \zeta_0)$$

$$+\{f(\xi_0, \eta, \zeta) - f(\xi_0, \eta_0, \zeta)\}$$

$$= \{f(\xi_0, \eta_0, \zeta) - f(\xi_0, \eta_0, \zeta_0)\}$$

In each bracket on the right hand side, we observe that only one of the independent variables changes its value. Hence we can apply to each bracket the mean value theorem for functions of a single variable and obtain

$$F(x) - F(x_0) = (\xi - \xi_0) f_\xi(\bar{\xi}, \eta, \xi) + (\eta - \eta_0) f_\eta(\xi_0, \bar{\eta}, \zeta)$$

$$+(\xi - \xi_0) f_\zeta(\xi_0, \eta_0, \bar{\zeta}),$$

where $\bar{\xi}$ lies between x and x_0, $\bar{\eta}$ between h_0 and h, and $\bar{\zeta}$ between z_0 and z. Moreover, by the mean value theorem, we have

$$\xi - \xi_0 = \phi(x) - \phi(x_0) = (x - x_0)\phi'(x_1),$$

$$\eta - \eta_0 = \psi(x) - \psi(x_0) = (x - x_0)\psi'(x_2),$$

$$\zeta - \zeta_0 = \chi(x) - \chi(x_0) = (x - x_0)\chi'(x_3),$$

where all x_1, x_2, and x_3 lie between x_0 and x. Substituting these values in the last equation and dividing by $x - x_0$, we have

$$\frac{F(x) - F(x_0)}{x - x_0} = f_\xi(\bar{\xi}, \eta, \zeta)\phi'(x_1) + f_\eta(\xi_0, \bar{\eta}, \zeta)\psi'(x_2)$$

$$+ f_\zeta(\xi_0, \eta_0, \bar{\zeta})\chi'(x_3)$$

We now let x tend to x_0. Owing to the continuity of $\phi(x)$, $\psi(x)$, $\chi(x)$, the quantities ξ,η,ζ tend to ξ_0,η_0,ζ_0, respectively, and, *a fortiori*, so do $\bar{\xi},\bar{\eta},\bar{\zeta}$. Also x_1, x_2 and x_3 tend to x_0. Since all the functions on the right hand side are continuous, we have

$$\lim_{n\to x_0}\frac{F(x)-F(x_0)}{x-x_0}=F'(x_0)=f_\xi(\xi_0,\eta_0,\zeta_0)\phi'(x_0)$$

$$+f_\eta(\xi_0,\eta_0,\zeta_0)\psi'(x_0)\ +f_\zeta(\xi_0,\eta_0,\zeta_0)\chi'(x_0),$$

thus establishing the formula for $F'(x)$.

The continuity of $F'(x)$ follows immediately from the formula, since, by assumption, ϕ', ψ', χ', are continuous f_ξ, f_η and f_ζ are are continuous functions of continuous functions. This theorem may be extended to compound functions of two or more variables, as follows:

If the funding $u=f(\xi,\eta,\ldots)$ has continuous partial derivatives of the first order in the region S and the functions $\xi=\phi(x,y)$, $\eta=\psi(x,y)$, x,y have continuous partial derivatives of the first order in R, then $u=F(x,y)=f\{\phi(x,y),\ \psi(x,y),\ldots$ has continuous partial derivatives of the first order in R and their derivatives are given by

$$F_x=f_\xi\phi_x+f_\eta\psi_x+\ldots,$$

$$F_y=f_\xi\phi_y+f_\eta\psi_y+\ldots.$$

These formulae are often written in the abbreviated form

$$u_x=u_\xi\xi_x+u_\eta\psi_x+\ldots,$$

$$u_y=u_\xi\xi_y+u_\eta\eta_y+\ldots.$$

In order to derive them, we temporarily introduce the notation $g(x)=\phi(x,y_0)$, $h(x)=\psi(x,y_0),\ldots$, where y_0 is a fixed value of y. By the definition of the partial derivatives, it follows that $g'(\mathrm{x})=\phi_x(x,y_0)$, $h'(\mathrm{x})=\psi_x(x,y_0)$,. Similarly, if we writs $H(\mathrm{x})=F(x,y_0)$, we have $H'(x)=F_x(x,y_0)$. We now apply the theorem just proved to the function $u=H(x)=f(\xi,\eta,\ldots)=f\{g(x),\ h(x),\ldots\}$ and obtain

$$H'(x_0) = f_\xi g'(x_0) + f_\eta h'(x_0) + ...$$

Returning to the original symbols, we have

$$F_x(x_0, y_0) = f_\xi \phi_x(x_0, y_0) + f_\eta \psi_x(x_0, y_0) + ...$$

The other formula is proved in a similar manner.

If we wish to calculate higher order derivatives, we need only differentiate again the right hand side of our formulae with respect to x and y, regarding $f_\xi, f_\eta, ...$ as compound functions. Thus, for u = $f(\xi, \eta) = f\{\phi(x,y), \psi(x,y)\}$, we have

$$u_{xx} = f_{\xi\xi}\phi_x^2 + 2f_{\xi\eta}\phi_x\psi_x + f_{\eta\eta}\psi_x^2 + f_\xi\phi_{xx} + f_\eta\psi_{xx},$$

$$u_{xx} = f_{\xi\xi}\phi_x\phi_y + f_{\xi\eta}(\phi_x\psi_y + \phi_y\psi_x) + f_{\eta\eta}\psi_x\psi_y + f_\xi\phi_{xy} + f_\eta\psi_{xy},$$

$$u_{yy} = f_{\xi\xi}\phi_y^2 + 2f_{\xi\eta}\phi_y\psi_y + f_{\eta\eta}\psi_y^2 + f_\xi\phi_{yy} + f_\eta\psi_{yy}.$$

Change of the Independent Variables

A particularly important type of compound function occurs in the process of changing the independent variables. For example, let $u = f(\xi, \eta)$ be a function of ξ and η, which we interpret as rectangular co-ordinates in the $\xi\eta$-plane. If we rotate the axes in the $\xi\eta$-planes through an angle θ, we obtain a new system of co-ordinates x,y, related to the co-ordinates (ξ, η) by the equations:

$$\xi = x\cos\theta - yx\sin\theta,\ \eta = x\sin\theta + y\cos\theta, \text{ or}$$

$$x = \xi\cos\theta + \eta\sin\theta,\ y = x\sin\theta + y\cos\theta, \text{ or}$$

The function $u = f(\xi, \eta)$ can then be expressed as a function of the new variables x, y:

$$u = f(\xi, \eta) = f(x, y).$$

The chain rule yields then immediately:

$$u_x = u_\xi\cos\theta + u_\eta\sin\theta,\ u_v = -u_\xi\sin\theta + u_\eta\cos\theta$$

Thus, the partial derivatives are transformed by the same formulae as the independent variables. This is true for rotation of the axes in space also. Another important type of change of co-ordinates is the change from rectangular co-ordinates x, y to polar co-ordinates r, θ.

This is done by means of the equations,

$$x = r\cos\theta,$$

$$y = r\sin\theta,$$

$$r = \sqrt{(x^2 + y^2)}$$

$$\theta = \arctan\frac{y}{x}$$

We then find that for an arbitrary function $u = f(x,x)$ with continuous partial derivatives of the first order we have

$$u = f(x, y) = f(r\cos\theta, r\sin\theta) = F(r, \theta),$$

$$u_x = u_r r_x + u_\theta \theta_x = u_r \frac{x}{r} - u_\theta \frac{y}{r^2} = u_r \cos\theta - u_\theta \frac{\sin\theta}{r},$$

$$u_y = u_r r_y + u_\theta \theta_y = u_r \frac{y}{r} - u_\theta \frac{x}{r^2} = u_r \sin\theta - u_\theta \frac{\cos\theta}{r}.$$

Whence we obtain the often useful equation

$$u_x^2 + u_r^2 = u_r^2 + \frac{1}{r^2} u_\theta^2$$

In general, let us consider a pair of functions $\xi = \phi(x, y)$, $\eta = \psi(x, y)$ which are continuous and have continuous derivatives in a region R of the xy-plane. These equations assign to each point (x, y) in R a point $\xi = \phi(x, y)$, $\eta = \psi(x, y)$ in the $\xi\eta$-plane. As (x, y) ranges over R, the corresponding point (ξ, η) will range over some set of values S in the $\xi\eta$-plane. Naturally, it is possible that several distinct points (x, y) will give the same values for (ξ, η), so that there corre-spond to several points (x, y) only one point (ξ, η).

We shall assume that this is not s, but instead that there corresponds to one point $Q(\xi, \eta)$ in S exactly one point $P(x,y)$ in R. We may therefore look at the correspondence from either point of view - saying that Q corresponds to P or that P corresponds to Q. The latter point of view can be expressed as follows: There corresponds to each point (ξ, η) in S one x and one y, namely, the co-ordinates of P, or, in, equations, there are two functions $x = g(\xi, \eta)$, $y = h(\xi, \eta)$, defined in S, which represent the corresponding inverse to $\xi = \phi(x,y)$, $\eta = \psi(x,y)$.

It happens often that the functions $g(\xi, \eta)$, $h(\xi, \eta)$ are by no means easy to calculate, even when they do exist, whence we shall now discover how to obtain the partial derivatives g_ξ, g_η, h_ξ, h_η directly from the partial derivatives ϕ_x, ϕ_y, ψ_x, ψ_y without at all calculating g and h themselves.

For this purpose, we observe that, if we choose any point $Q(\xi, \eta)$, find the corresponding point $P\{g(\xi,\eta),h(\xi,\eta)\}$ in R and then find the point in S corresponding to P, which is $\phi\{g(\xi,\eta),h(\xi,\eta)\}$, $\psi\{g(\xi,\eta),h(\xi,\eta)\}$, we have simply returned to the point Q. In other words, the equations $\xi = \phi\{g(\xi,\eta),h(\xi,\eta)\}$, $\eta = \psi\{g(\xi,\eta),h(\xi,\eta)\}$ are identities in ξ and η. We now differentiate both sides of both equations with respect to ξ and η and find

$$1 = \phi_x g_\xi + \phi_y h_\xi, \quad 0 = \phi_x g_\eta + \phi_y h_\eta,$$

$$0 = \psi_x g_\xi + \psi_y h_\xi, \quad 1 = \psi_x g_\eta + \psi_y h_\eta,$$

If an equation expresses an identical relationship, differentiation with respect to any independent variable in it yields an identity, as follows immediately from the definition.

By solving these equations, we find that

$$g_\xi = \frac{\psi_y}{D}, g_n = -\frac{\phi_y}{D}, h_\xi = -\frac{\psi_x}{D}, h_\eta = \frac{\phi_x}{D}$$

or

$$g_\xi = \frac{\eta_y}{D}, x_n = -\frac{\xi_y}{D}, y_\xi = -\frac{\eta_x}{D}, y_\eta = \frac{\xi_x}{D}$$

where by D is the determinant

$$D = \xi x \eta_v - \xi_y \eta_x = \begin{vmatrix} \dfrac{\partial \xi}{\partial x} & \dfrac{\partial \xi}{\partial y} \\ \dfrac{\partial \eta}{\partial x} & \dfrac{\partial \eta}{\partial y} \end{vmatrix}$$

which we assume to be non-zero. This determinant D, called the functional determinant or Jacobian of (ξ, η) with respect to (x,y), occurs so frequently that one uses frequently the special symbol

$$D = \frac{\partial(\xi, \eta)}{\partial(x,y)}$$

Implicit Functions

In our study of functions of several variables, we have as yet had no analogue to the inverse function. We can regard the inverse function of $y = f(x)$ to be the function obtained by solving the equation $y - f(x) = 0$ for x. In this section, we shall seek more generally to solve equations $F(x, y) = 0$ for x or y and to discuss functions of several variables in a corresponding way. Even in elementary analytical geometry, curves are frequently represented not by equations $y = f(\text{x})$ or $x = \phi(\text{y})$, but by an equation involving x and y in the form $F(x, y) = 0$. For example, we have the circle $x + y - 1 = 0$, the ellipse $x/a + y/b - 1 = 0$ and. the lemniscate $(x + y) - 2a(x - y) = 0$. In order to obtain y as a function of x or x as a function of y, we must solve the equation for y or for x. We then say that the function $y = f(x)$ or $x = \phi\,(y)$ found in this way is defined implicitly by the equation $F(x, y) = 0$, and that the solution of this equation yields the function explicitly. In the examples cited and in many others, the solution can be carried out and the solutions stated explicitly in terms of the elementary functions. In other cases, the solution can be obtained in terms of an infinite series or other limiting processes, i.e., we can approximate to the solution $y = f(x)$ or $x = \phi(y)$ as closely as we please. However, for many purposes, it is more convenient to base the discussion on the implicit definition $F(x, y) = 0$, instead of resorting to an exact or approximate solution of the equation. The idea that every function $F(x, y)$ yields a function $y = f(x)$ or $x = \phi(y)$ given implicitly by means of the equation $F(x, y) = 0$ is erroneous. On the contrary, it is easy to give examples of functions $F(x, y)$ which, when equated to zero, permit no solution in terms of functions of one variable. Thus, for example, the equation $x + y = 0$ is satisfied by the single pair of values $x = 0$, $y = 0$ only, while the equation $x + y + 1 = 0$ is not satisfied by any real values. It is therefore necessary to investigate this matter more closely, in order to find out whether an equation $F(x, y) = 0$ can actually be solved.

Chapter 5

Integral Methods and Design

Integration is an important concept in mathematics and, together with its inverse, differentiation, is one of the two main operationsin calculus. Given a function f of a real variable x and an interval$[a, b]$ of the real line, the definite integral

$$\int_a^b f(x)dx$$

is defined informally to be the area of the region in the xy-plane bounded by the graph of f, the x-axis, and the vertical lines $x = a$ and $x = b$, such that areas above the axis add to the total, and the area below the x axis subtract from the total.

The term *integral* may also refer to the notion of antiderivative, a function F whose derivative is the given function f. In this case, it is called an *indefinite integral* and is written:

$$F = \int f(x)dx.$$

The integrals discussed in this article are termed *definite integrals*.

The principles of integration were formulated independently by Isaac Newton and Gottfried Leibniz in the late 17th century.

Through the fundamental theorem of calculus, which they independently developed, integration is connected with differentiation: if f is a continuous real-valued function defined on a closed interval $[a, b]$, then, once an antiderivative F of f is known, the definite integral of f over that interval is given by

$$\int_a^b f(x)dx = F(b) - F(a)$$

Integrals and derivatives became the basic tools of calculus, with numerous applications in science and engineering. The founders of the calculus thought of the integral as an infinite sum of rectangles of infinitesimal width. A rigorous mathematical definition of the integral was given by Bernhard Riemann. It is based on a limiting procedure which approximates the area of a curvilinear region by breaking the region into thin vertical slabs. Beginning in the nineteenth century, more sophisticated notions of integrals began to appear, where the type of the function as well as the domain over which the integration is performed has been generalised. A line integral is defined for functions of two or three variables, and the interval of integration $[a, b]$ is replaced by a certain curve connecting two points on the plane or in the space. In a surface integral, the curve is replaced by a piece of a surface in the three-dimensional space. Integrals of differential forms play a fundamental role in modern differential geometry. These generalizations of integrals first arose from the needs of physics, and they play an important role in the formulation of many physical laws, notably those of electrodynamics. There are many modern concepts of integration, among these, the most common is based on the abstract mathematical theory known as Lebesgue integration, developed by Henri Lebesgue.

Pre-calculus Integration

The first documented systematic technique capable of determining integrals is the method of exhaustion of the ancient Greek astronomer Eudoxus (*ca.* 370 BC), which sought to find areas and volumes by breaking them up into an infinite number of shapes for which the area or volume was known. This method was further developed and employed by Archimedes in the 3rd century BC and used to calculate areas for parabolas and an approximation to the area of a circle. Similar methods were independently developed in China around the 3rd century AD by Liu Hui, who used it to find the area of the circle. This method was later used in the 5th century by Chinese father-and-son mathematicians Zu Chongzhi and Zu Geng to find the volume of a sphere (Shea 2007; Katz 2004, pp. 125–126).

The next significant advances in integral calculus did not begin to appear until the 16th century. At this time the work of Cavalieri with his *method of indivisibles*, and work by Fermat, began to lay the foundations of modern calculus, with Cavalieri computing the integrals of x^n up to degree $n = 9$ in Cavalieri's quadrature formula. Further steps were made in the early 17th century by Barrow and Torricelli, who provided the first hints of a connection between integration and differentiation. Barrow provided the first proof of the fundamental theorem of calculus. Wallis generalized Cavalieri's method, computing integrals of x to a general power, including negative powers and fractional powers.

Newton and Leibniz

The major advance in integration came in the 17th century with the independent discovery of the fundamental theorem of calculus by Newton and Leibniz. The theorem demonstrates a connection between integration and differentiation. This connection, combined with the comparative ease of differentiation, can be exploited to calculate integrals.

In particular, the fundamental theorem of calculus allows one to solve a much broader class of problems. Equal in importance is the comprehensive mathematical framework that both Newton and Leibniz developed. Given the name infinitesimal calculus, it allowed for precise analysis of functions within continuous domains. This framework eventually became modern calculus, whose notation for integrals is drawn directly from the work of Leibniz.

Formalizing Integrals

While Newton and Leibniz provided a systematic approach to integration, their work lacked a degree of rigour. Bishop Berkeley memorably attacked the vanishing increments used by Newton, calling them "ghosts of departed quantities". Calculus acquired a firmer footing with the development of limits. Integration was first rigorously formalized, using limits, by Riemann.

Although all bounded piecewise continuous functions are Riemann integrable on a bounded interval, subsequently more general functions were considered – particularly in the context of Fourier analysis – to which Riemann's definition does not apply, and Lebesgue formulated a different definition of integral, founded in measure theory (a subfield of real analysis). Other definitions of integral, extending Riemann's and Lebesgue's approaches, were proposed. These approaches based on the real number system are the ones most common today, but alternative approaches exist, such as a definition of integral as the standard part of an infinite Riemann sum, based on the hyperreal number system.

Historical Notation

Isaac Newton used a small vertical bar above a variable to indicate integration, or placed the variable inside a box. The vertical bar was easily confused with $\dot{x}$ or x' , which Newton used to indicate differentiation, and the box notation was difficult for printers to reproduce, so these notations were not widely adopted.

The modern notation for the indefinite integral was introduced by Gottfried Leibniz in 1675 (Burton 1988, p. 359; Leibniz 1899, p. 154). He adapted the integral symbol, *f*, from the letter *f* (long s), standing for *summa* (written as *numma*; Latin for "sum" or "total"). The modern notation for the definite integral, with limits above and below the integral sign, was first used by Joseph Fourier in *Mémoires* of the French Academy around 1819–20, reprinted in his book of 1822.

Terminology and Notation

The simplest case, the integral over x of a real-valued function $f(x)$, is written as

$$\int f(x)dx.$$

The integral sign *f* represents integration. The dx indicates that we are integrating over x; dx is called the variable of integration. In correct mathematical typography, the dx is separated from the integrand by a space (as shown). Some authors use an upright d (that is, dx instead of dx). Inside the *f*...dx is the expression to be integrated, called the *integrand*. In this case the integrand is the function $f(x)$. Because there is no domain specified, the integral is called an *indefinite integral*.

When integrating over a specified domain, we speak of a *definite integral.* Integrating over a domain D is written as $\int_D f(x)dx$, or $\int_a^b f(x)dx$ if the domain is an interval $[a, b]$ of x;

The domain D or the interval $[a, b]$ is called the *domain of integration.*

If a function has an integral, it is said to be *integrable.* In general, the integrand may be a function of more than one variable, and the domain of integration may be an area, volume, a higher dimensional region, or even an abstract space that does not have a geometric structure in any usual sense (such as a sample space in probability theory).

In the modern Arabic mathematical notation, which aims at pre-university levels of education in the Arab world and is written from right to left, a reflected integral symbol ∫ is used (W3C 2006).

The variable of integration dx has different interpretations depending on the theory being used. It can be seen as strictly a notation indicating that x is a dummy variable of integration; if the integral is seen as a Riemann sum, dx is a reflection of the weights or widths d of the intervals of x; in Lebesgue integration and its extensions, dx is a measure; in non-standard analysis, it is an infinitesimal; or it can be seen as an independent mathematical quantity, a differential form. More complicated cases may vary the notation slightly.

In Leibniz's notation, dx is interpreted an infinitesimal change in x, but his interpretation lacks rigour in the end. Nonetheless Leibniz's notation is the most common one today; and as few people are in need of full rigour, even his interpretation is still used in many settings.

Integrals appear in many practical situations. If a swimming pool is rectangular with a flat bottom, then from its length, width, and depth we can easily determine the volume of water it can contain (to fill it), the area of its surface (to cover it), and the length of its edge (to rope it). But if it is oval with a rounded bottom, all of these quantities call for integrals. Practical approximations may suffice for such trivial examples, but precision engineering (of any discipline) requires exact and rigorous values for these elements.

Approximations to integral of √x from 0 to 1, with % 5 right samples (above) and % 12 left samples (below)

To start off, consider the curve $y = f(x)$ between $x = 0$ and $x = 1$ with $f(x) = \sqrt{x}$. We ask:

What is the area under the function f, in the interval from 0 to 1?

and call this (yet unknown) area the integral of f. The notation for this integral will be

$$\int_0^1 \sqrt{x}\,dx.$$

As a first approximation, look at the unit square given by the sides $x = 0$ to $x = 1$ and $y = f(0) = 0$ and $y = f(1) = 1$. Its area is exactly 1. As it is, the true value of the integral must be somewhat less. Decreasing the width of the approximation rectangles shall give a better result; so cross the interval in five steps, using the approximation points 0, 1/5, 2/5, and so on to 1. Fit a box for each step using the right end height of each curve piece, thus √(1D†5), √(2D†5), and so on to √1 = 1. Summing the areas of these rectangles, we get a better approximation for the sought integral, namely

$$\sqrt{\frac{1}{5}}\left(\frac{1}{5}-0\right)+\sqrt{\frac{2}{5}}\left(\frac{2}{5}-\frac{1}{5}\right)+\cdots+\sqrt{\frac{5}{5}}\left(\frac{5}{5}-\frac{4}{5}\right)\approx 0.7497.$$

Notice that we are taking a sum of finitely many function values of f, multiplied with the differences of two subsequent approximation points. We can easily see that the approximation is still too large. Using more steps produces a closer approximation, but will never be exact: replacing the 5 subintervals by twelve as depicted, we will get an approximate value for the area of 0.6203, which is too small. The key idea is the transition from adding *finitely many* differences of approximation points multiplied by their respective function values to using infinitely many fine, or *infinitesimal* steps.

As for the *actual calculation of integrals*, the fundamental theorem of calculus, due to Newton and Leibniz, is the fundamental link between the operations of differentiating and integrating. Applied to the square root curve, $f(x) = x^{1/2}$, it says to look at the antiderivative $F(x) = (2/3)x^{3/2}$, and simply take $F(1) - F(0)$, where 0 and 1 are the boundaries of the interval [0,1]. So the *exact* value of the area under the curve is computed formally as

$$\int_0^1 \sqrt{x}\,dx = \int_0^1 x^{1/2}\,dx = F(1) - F(0) = 2/3.$$

(This is a case of a general rule, that for $f(x) = x^q$, with $q \neq -1$, the related function, the so-called antiderivative is $F(x) = x^{q+1}/(q + 1)$.)

The notation

$$\int f(x)dx$$

conceives the integral as a weighted sum, denoted by the elongated *s*, of function values, $f(x)$, multiplied by infinitesimal step widths, the so-called *differentials*, denoted by dx. The multiplication sign is usually omitted.

Historically, after the failure of early efforts to rigorously interpret infinitesimals, Riemann formally defined integrals as a limit of weighted sums, so that the dx suggested the limit of a difference (namely, the interval width). Shortcomings of Riemann's dependence on intervals and continuity motivated newer definitions, especially the Lebesgue integral, which is founded on an ability to extend the idea of "measure" in much more flexible ways. Thus the notation

$$\int_A f(x)d\mu$$

refers to a weighted sum in which the function values are partitioned, with μ measuring the weight to be assigned to each value. Here A denotes the region of integration.

Differential geometry, with its "calculus on manifolds", gives the familiar notation yet another interpretation. Now $f(x)$ and dx become a differential form, $\omega = f(x)\,dx$, a new differential operator d, known as the exterior derivative is introduced, and the fundamental theorem becomes the more general Stokes' theorem,

$$\int_A d\omega = \int_{\partial A} \omega,$$

from which Green's theorem, the divergence theorem, and the fundamental theorem of calculus follow.

More recently, infinitesimals have reappeared with rigor, through modern innovations such as non-standard analysis. Not only do these methods vindicate the intuitions of the pioneers; they also lead to new mathematics. Although there are differences between these conceptions of integral, there is considerable overlap. Thus, the area of the surface of the oval swimming pool can be handled as a geometric ellipse, a sum of infinitesimals, a Riemann integral, a Lebesgue integral, or as a manifold with a differential form. The calculated result will be the same for all.

Formal Definitions

There are many ways of formally defining an integral, not all of which are equivalent. The differences exist mostly to deal with differing special cases which may not be integrable under other definitions, but also occasionally for pedagogical reasons. The most commonly used definitions of integral are Riemann integrals and Lebesgue integrals.

Other Integrals

Riemann–Stieltjes Integral

In mathematics, the Riemann–Stieltjes integral is a generalization of the Riemann integral, named after Bernhard Riemann and Thomas Joannes Stieltjes.

Definition

The Riemann–Stieltjes integral of a real-valued function f of a real variable with respect to a real function g is denoted by

$$\int_a^b f(x)dg(x)$$

and defined to be the limit, as the mesh of the partition

$$P = \{a = x_0 < x_1 < \cdots < x_n = b\}$$

of the interval $[a, b]$ approaches zero, of the approximating sum

$$S(P,f,g)=\sum_{i=0}^{n-1} f(c_i)(g(x_{i+1})-g(x_i))$$

where c_i is in the i-th subinterval $[x_i, x_{i+1}]$. The two functions f and g are respectively called the integrand and the integrator.

The "limit" is here understood to be a number A (the value of the Riemann–Stieltjes integral) such that for every $\varepsilon > 0$, there exists $\delta > 0$ such that for every partition P with mesh(P) < δ, and for every choice of points c_i in $[x_i, x_{i+1}]$,

$$|S(P,f,g)-A|<\varepsilon.$$

Generalized Riemann–Stieltjes Integral

A slight generalization, introduced by Pollard (1920) and now standard in analysis, is to consider in the above definition partitions P that *refine* another partition P_ε, meaning that P arises from P_ε by the addition of points, rather than from partitions with a finer mesh. Specifically, the generalized Riemann–Stieltjes integral of f with respect to g is a number A such that for every $\varepsilon > 0$ there exists a partition P_ε such that for every partition P that refines P_ε,

$$|S(P,f,g)-A|<\varepsilon$$

for every choice of points c_i in $[x_i, x_{i+1}]$.

This generalization exhibits the Riemann–Stieltjes integral as the Moore–Smith limit on the directed set of partitions of $[a, b]$ (McShane 1952). Hildebrandt (1938) calls it the Pollard–Moore–Stieltjes integral.

Darboux Sums

The Riemann–Stieltjes integral can be efficiently handled using an appropriate generalization of Darboux sums. For a partition P and a nondecreasing function g on $[a, b]$ define the upper Darboux sum of f with respect to g by

$$U(P,f,g)=\sum_{i=1}^{n}\sup_{x\in[x_i,x_{i+1}]} f(x)(g(x_{i+1})-g(x_i))$$

and the lower sum by

$$L(P,f,g)=\sum_{i=1}^{n}\inf_{x\in[x_i,x_{i+1}]} f(x)(g(x_{i+1})-g(x_i)).$$

Then the generalized Riemann–Stieltjes of f with respect to g exists if and only if, for every $\varepsilon > 0$, there exists a partition P such that

$$U(P, f, g) - L(P, f, g) < \varepsilon.$$

Furthermore, f is Riemann–Stieltjes integrable with respect to g (in the classical sense) if

$$\lim_{\text{mesh}(P)\to 0} [U(P, f, g) - L(P, f, g)] = 0.$$

Properties and Relation to the Riemann Integral

If g should happen to be everywhere differentiable, then the Riemann–Stieltjes integral may still be different from the Riemann integral of $f(x)g'(x)$ given by

$$\int_a^b f(x)g'(x)dx,$$

for example, if the derivative is unbounded. But if the derivative is continuous, they will be the same. This condition is also satisfied if g is the (Lebesgue) integral of its derivative; in this case g is said to be absolutely continuous.

However, g may have jump discontinuities, or may have derivative zero *almost* everywhere while still being continuous and increasing (for example, g could be the Cantor function), in either of which cases the Riemann–Stieltjes integral is not captured by any expression involving derivatives of g.

The Riemann–Stieltjes integral admits integration by parts in the form

$$\int_a^b f(x)dg(x) = f(b)g(b) - f(a)g(a) - \int_a^b g(x)df(x).$$

and the existence of either integral implies the existence of the other (Hille & Phillips 1974, §3.3).

Existence of the Integral

The best simple existence theorem states that if f is continuous and g is of bounded variation on $[a, b]$, then the integral exists. A function g is of bounded variation if and only if it is the difference between two monotone functions. If g is not of bounded variation, then there will be continuous functions which cannot be integrated with respect to g. In general, the integral is not well-defined if f and g share any points of discontinuity, but this sufficient condition is not necessary.

On the other hand, a classical result of Young (1936) states that the integral is well-defined if f is α-Hölder continuous and g is β-Hölder continuous with $\alpha + \beta > 1$.

Application to Probability Theory

If g is the cumulative probability distribution function of a random variable X that has a probability density function with respect to Lebesgue measure, and f is any function for which the expected value $E(|f(X)|)$ is finite, then the probability density function of X is the derivative of g and we have

$$E(f(X)) = \int_{-\infty}^{\infty} f(x)g'(x)dx.$$

But this formula does not work if X does not have a probability density function with respect to Lebesgue measure. In particular, it does not work if the distribution of X is discrete (i.e., all of the probability is accounted for by point-masses), and even if the cumulative distribution function g is continuous, it does not work if g fails to be absolutely continuous (again, the Cantor function may serve as an example of this failure). But the identity

$$E(f(X)) = \int_{-\infty}^{\infty} f(x)dg(x)$$

holds if g is *any* cumulative probability distribution function on the real line, no matter how ill-behaved. In particular, no matter how ill-behaved the cumulative distribution function g of a random variable X, if the moment $E(X^n)$ exists, then it is equal to

$$E(X^n) = \int_{-\infty}^{\infty} x^n dg(x).$$

Application to Functional Analysis

The Riemann–Stieltjes integral appears in the original formulation of F. Riesz's theorem which represents the dual space of the Banach space $C[a,b]$ of continuous functions in an interval $[a,b]$ as Riemann–Stieltjes integrals against functions of bounded variation. Later, that theorem was reformulated in terms of measures.

The Riemann–Stieltjes integral also appears in the formulation of the spectral theorem for (non-compact) self-adjoint (or more generally, normal) operators in a Hilbert space. In this theorem, the integral is considered with respect to a spectral family of projections.

Generalization

An important generalization is the Lebesgue–Stieltjes integral which generalizes the Riemann–Stieltjes integral in a way analogous to how the Lebesgue integral generalizes the Riemann integral. If improper Riemann–Stieltjes integrals are allowed, the Lebesgue integral is not strictly more general than the Riemann–Stieltjes integral.

The Riemann–Stieltjes integral also generalizes to the case when either the integrand f or the integrator g take values in a Banach space. If $g : [a,b] \to X$ takes values in the Banach space X, then it is natural to assume that it is of strongly bounded variation, meaning that

$$\sup \sum_i \| g(t_i) - g(t_{i+1}) \|_X < \infty$$

the supremum being taken over all finite partitions

$$a = t_0 \le t_1 \le \cdots \le t_n = b$$

of the interval $[a,b]$. This generalization plays a role in the study of semigroups, via the Laplace–Stieltjes transform.

Lebesgue–Stieltjes Integration

In measure-theoretic analysis and related branches of mathematics, Lebesgue–Stieltjes integration generalizes Riemann–Stieltjes and Lebesgue integration, preserving the many advantages of the former in a more general measure-theoretic framework.

The Lebesgue–Stieltjes integral is the ordinary Lebesgue integral with respect to a measure known as the Lebesgue–Stieltjes measure, which may be associated to any function of bounded variation on the real line. The Lebesgue–Stieltjes measure is a regular Borel measure, and conversely every regular Borel measure on the real line is of this kind.

Lebesgue–Stieltjes integrals, named for Henri Leon Lebesgue and Thomas Joannes Stieltjes, are also known as Lebesgue–Radon integrals or just Radon integrals, after Johann Radon, to whom much of the theory is due. They find common application in probability and stochastic processes, and in certain branches of analysis including potential theory.

Definition

The Lebesgue–Stieltjes integral

$$\int_a^b f(x)dg(x)$$

is defined when $f : [a,b] \to \mathrm{R}$ is Borel-measurable and bounded and $g : [a,b] \to \mathrm{R}$ is of bounded variation in $[a,b]$ and right-continuous, or when f is non-negative and g is monotone and right-continuous. To start, assume that f is non-negative and g is monotone non-decreasing and right-continuous. Define $w((s,t]) := g(t) - g(s)$ and $w(\{a\}) := 0$ (Alternatively, the construction works for g left-continuous, $w([s,t)) := g(t) - g(s)$ and $w(\{b\}) := 0$).

By Carathéodory's extension theorem, there is a unique Borel measure μ_g on $[a,b]$ which agrees with w on every interval I. The measure μ_g arises from an outer measure (in fact, a metric outer measure) given by

$$\mu_g(E) = \inf\left\{\sum_i \mu_g(I_i) \mid E \subset \bigcup_i I_i\right\}$$

the infimum taken over all coverings of E by countably many semiopen intervals. This measure is sometimes called the Lebesgue–Stieltjes measure associated with g.

The Lebesgue–Stieltjes integral

$$\int_a^b f(x)dg(x)$$

is defined as the Lebesgue integral of f with respect to the measure μ_g in the usual way. If g is non-increasing, then define

$$\int_a^b f(x)dg(x) := -\int_a^b f(x)d(-g)(x),$$

the latter integral being defined by the preceding construction. If g is of bounded variation and f is bounded, then it is possible to write

$$g(x) = g_1(x) - g_2(x)$$

where $g_1(x) := Vx$

ag is the total variation of g in the interval $[a,x]$, and $g_2(x) = g_1(x) - g(x)$. Both g_1 and g_2 are monotone non-decreasing. Now the Lebesgue–Stieltjes integral with respect to g is defined by

$$\int_a^b f(x)dg(x) = \int_a^b f(x)dg_1(x) - \int_a^b f(x)dg_2(x),$$

where the latter two integrals are well-defined by the preceding construction.

Daniell Integral

An alternative approach (Hewitt & Stromberg 1965) is to define the Lebesgue–Stieltjes integral as the Daniell integral that extends the usual Riemann–Stieltjes integral. Let g be a non-increasing right-continuous function on $[a,b]$, and define $I(f)$ to be the Riemann–Stieltjes integral

$$I(f) = \int_a^b f(x)dg(x)$$

for all continuous functions f. The functional I defines a Radon measure on $[a,b]$. This functional can then be extended to the class of all non-negative functions by setting

$$\overline{I}(h) = \sup\{I(f) \mid f \in C[a,b], 0 \le f \le h\}$$

and

$$\overline{\overline{I}}(h) = \inf\{I(f) \mid f \in C[a,b], h \le f\}.$$

For Borel measurable functions, one has

$$\overline{I}(h) = \overline{\overline{I}}(h),$$

and either side of the identity then defines the Lebesgue–Stieltjes integral of h. The outer measure μ_g is defined via

$$\mu_g(A) = \overline{\overline{I}}(\chi_A)$$

where χ_A is the indicator function of A.

Integrators of bounded variation are handled as above by decomposing into positive and negative variations.

Example

Suppose that $\gamma:[a,b]\to\mathbb{R}^2$ is a rectifiable curve in the plane and $\rho:\mathbb{R}^2\to[0,\infty)$ is Borel measurable. Then we may define the length of γ with respect to the Euclidean metric weighted by ρ to be $\int_a^b \rho(\gamma(t))d\ell(t)$, where $\ell(t)$ is the length of the restriction of γto $[a,t]$. This is sometimes called the ρ-length of γ. This notion is quite useful for various applications: for example, in muddy terrain the speed in which a person can move may depend on how deep the mud is. If $\rho(z)$ denotes the inverse of the walking speed at or near z, then the ρ-length of γis the time it would take to traverse γ. The concept of extremal length uses this notion of the ρ-length of curves and is useful in the study of conformal mappings.

Integration by Parts

A function *f* is said to be "regular" at a point if the right $f(a+)$ and left hand limits and $f(a-)$ exist, and the function takes the average value,

$$f(a)=\frac{1}{2}\big(f(a-)+f(a+)\big),$$

at the limiting point. Given two functions *U* and *V* of finite variation, if at each point either *U* or *V* is continuous, or if both *U* and *V* are regular, then there is an integration by parts formula for the Lebesgue–Stieltjes integral:

$$\int_a^b U\,dV+\int_a^b V\,dU=U(b+)V(b+)-U(a-)V(a-),$$

where $b>a$. Under a slight generalization of this formula, the extra conditions on *U* and *V* can be dropped. An alternative result, of significant importance in the theory of Stochastic calculus is the following. Given two functions *U* and *V* of finite variation, which are both right-continuous and have left-limits (they are cadlag functions) then

$$U(t)V(t) = U(0)V(0) + \int_{(0,t]} U\,dV + \int_{(0,t]} V\,dU + \sum_{u\in(0,t]} \Delta U_u \Delta V_u,$$

where $\Delta U_u = U(t) - U(t-)$. This result can be seen as a precursor to Ito's lemma, and is of use in the general theory of Stochastic integration. The final term is $\Delta U(t)\Delta V(t) = d[U,V]$, which arises from the quadratic covariation of *U*and *V*.

Related Concepts

Lebesgue Integration: When $g(x) = x$ for all real x, then μ_g is the Lebesgue measure, and the Lebesgue–Stieltjes integral of f with respect to g is equivalent to the Lebesgue integral of f.

Riemann–Stieltjes Integration and Probability Theory

Where f is a continuous real-valued function of a real variable and v is a non-decreasing real function, the Lebesgue–Stieltjes integral is equivalent to the Riemann–Stieltjes integral, in which case we often write

$$\int_a^b f(x)dv(x)$$

for the Lebesgue–Stieltjes integral, letting the measure μ_v remain implicit. This is particularly common in probability theory when v is the cumulative distribution function of a real-valued random variable X, in which case

$$\int_{-\infty}^{\infty} f(x)dv(x) = \mathrm{E}[f(X)].$$

Daniell Integral

In mathematics, the Daniell integral is a type of integration that generalizes the concept of more elementary versions such as the Riemann integral to which students are typically first introduced.

One of the main difficulties with the traditional formulation of the Lebesgue integral is that it requires the initial development of a workable measure theory before any useful results for the integral can be obtained. However, an alternative approach is available, developed by Percy J. Daniell (1918) that does not suffer from this deficiency, and has a few significant advantages over the traditional formulation, especially as the integral is generalized into higher dimensional spaces and further generalizations such as the Stieltjes integral. The basic idea involves the axiomatization of the integral.

The Daniell Axioms

We start by choosing a family H of bounded real functions (called *elementary functions*) defined over some set X, that satisfies these two axioms:

1. H is a linear space with the usual operations of addition and scalar multiplication.
2. If a function $h(x)$ is in H, so is its absolute value $|h(x)|$.

In addition, every function h in H is assigned a real number Ih, which is called the *elementary integral* of h, satisfying these three axioms:

1. Linearity. If h and k are both in H, αand βand are any two real numbers, then $I(\alpha h+\beta k)=\alpha Ih+\beta Ik$.
2. Nonnegativity. If $h(x)\geq 0$, then $Ih\geq 0$.
3. Continuity. If $h_n(x)$ is a nonincreasing sequence (i.e. $h_1\geq\cdots\geq h_k\geq\cdots$) of functions in H that converges to 0 for all x in X, then $Ih_n\rightarrow 0$.

That is, we define a continuous non-negative linear functional over the space of elementary functions.

These elementary functions and their elementary integrals may be any set of functions and definitions of integrals over these functions which satisfy these axioms.

The family of all step functions evidently satisfies the above axioms for elementary functions. Defining the elementary integral of the family of step functions as the (signed) area underneath a step function evidently satisfies the given axioms for an elementary integral. Applying the construction of the Daniell integral described further below using step functions as elementary functions produces a definition of an integral equivalent to the Lebesgue integral.

Using the family of all continuous functions as the elementary functions and the traditional Riemann integral as the elementary integral is also possible, however, this will yield an integral that is also equivalent to Lebesgue's definition.

Doing the same, but using the Riemann–Stieltjes integral, along with an appropriate function of bounded variation, gives a definition of integral equivalent to the Lebesgue–Stieltjes integral.

Sets of measure zero may be defined in terms of elementary functions as follows. A set Z which is a subset of X is a set of measure zero if for any $\epsilon > 0$, there exists a nondecreasing sequence of nonnegative elementary functions $h_p(x)$ in H such that $Ih_p < \epsilon$ and $\sup_p h_p(x) \geq 1$ on Z.

A set is called a set of full measure if its complement, relative to X, is a set of measure zero. We say that if some property holds at every point of a set of full measure (or equivalently everywhere except on a set of measure zero), it holds almost everywhere.

Definition of the Daniell Integral

We can then proceed to define a larger class of functions, based on our chosen elementary functions, the class L^+, which is the family of all functions that are the limit of a nondecreasing sequence h_n of elementary functions almost everywhere, such that the set of integrals Ih_n is bounded. The integral of a function f in L^+ is defined as:

$$If = \lim_{n\to\infty} Ih_n$$

It can be shown that this definition of the integral is well-defined, i.e. it does not depend on the choice of sequence h_n.

However, the class L^+ is in general not closed under subtraction and scalar multiplication by negative numbers, but we can further extend it by defining a wider class of functions L such that every function $\phi(x)$ can be represented on a set of full measure as the difference $\phi = f - g$, for some functions *f* and g in the class L^+. Then the integral of a function $\phi(x)$ can be defined as:

$$\int_X \phi(x)dx = If - Ig$$

Again, it may be shown that this integral is well-defined, i.e. it does not depend on the decomposition of into *f* and g. This is the final construction of the Daniell integral.

Properties

Nearly all of the important theorems in the traditional theory of the Lebesgue integral, such as Lebesgue's dominated convergence theorem, the Riesz–Fischer theorem, Fatou's lemma, and Fubini's theorem may also readily be proved using this construction. Its properties are identical to the traditional Lebesgue integral.

Measures from the Daniell integral: Because of the natural correspondence between sets and functions, it is also possible to use the Daniell integral to construct a measure theory. If we take the characteristic function $\chi(x)$ of some set, then its integral may be taken as the measure of the set. This definition of measure based on the Daniell integral can be shown to be equivalent to the traditional Lebesgue measure.

Advantages Over the Traditional Formulation

This method of constructing the general integral has a few advantages over the traditional method of Lebesgue, particularly in the field of functional analysis. The Lebesgue and Daniell constructions are equivalent, as pointed out above, if ordinary finite-valued step functions are chosen as elementary functions. However, as one tries to extend the definition of the integral into more complex domains (e.g. attempting to define the integral of a linear functional), one runs into practical difficulties using

Lebesgue's construction that are alleviated with the Daniell approach. The Polish mathematician Jan Mikusinski has made an alternative and more natural formulation of Daniell integration by using the notion of absolutely convergent series. His formulation works for Bochner integral (Lebesgue integral for mappings taking values in Banach spaces). Mikusinski's lemma allows one to define integral without mentioning null sets. He also proved change of variables theorem for multiple integral for Bochner integrals and Fubini's theorem for Bochner integrals using Daniell integration. The book by Asplund and Bungart carries a lucid treatment of this approach for real valued functions. It also offers a proof of an abstract Radon–Nikodym theorem using Daniell–Mikusinski approach.

Haar Integral

Using the general theory of Lebesgue integration, one can then define an integral for all Borel measurable functions *f* on *G*. This integral is called the Haar integral. If μ is a left Haar measure, then

$$\int_G f(sx)\, d\mu(x) = \int_G f(x)\, d\mu(x)$$

for any integrable function *f*. This is immediate for characteristic functions, being essentially the definition of left invariance.

Examples

- A Haar measure on the topological group (R, +) which takes the value 1 on the interval [0,1] is equal to the restriction of Lebesgue measure to the Borel subsets of R. This can be generalized to (R^n, +).
- If *G* is the group of nonzero real numbers with multiplication as operation, then a Haar measure μ is given by

$$\mu(S) = \int_S \frac{1}{|t|} dt$$

for any Borel subset *S* of the nonzero reals.

This generalizes to the following:

For *G* = *GL*(n,R), any left Haar measure is a right Haar measure and one such measure μ is given by

$$\mu(S) = \int_S \frac{1}{|\det(X)|^n} dX$$

where dX denotes the Lebesgue measure on $\mathbb{R}n^2$, the set of all $n \times n$-matrices. This follows from the change of variables formula. More generally, on any Lie group of dimension d a left Haar measure can be associated with any non-zero left-invariant d-form ω, as the *Lebesgue measure* $|\omega|$; and similarly for right Haar measures.

This means also that the modular function can be computed, as the absolute value of the determinant of the adjoint representation. In order to define a Haar measure μ on the unit circle *T*, consider the function *f* from [0,2π] onto *T* defined by *f*(*t*) = (cos(*t*),sin(*t*)). Then μ can be defined by

$$\mu(S) = \frac{1}{2\pi} m\left(f^{-1}(S)\right),$$

where *m* is the Lebesgue measure. The factor $(2\pi)^{-1}$ is chosen so that μ(*T*) = 1.

If *G* is the group of non-null quaternions, then *G* can be seen as an open subset of R^4. A Haar measure μ is given by

$$\mu(S) = \int_S \frac{1}{(x^2 + y^2 + z^2 + w^2)^2} dx\,dy\,dz\,dw$$

where *dx dy dz dw* denotes the Lebesgue measure in R^4 and *S* is a Borel subset of G.

Henstock–Kurzweil Integral

In mathematics, the Henstock–Kurzweil integral, also known as the Denjoy integral (pronounced [dQÞÈ'wa]) and the Perron integral, is one of a number of definitions of the integral of a function. It is a generalization of the Riemann integral which in some situations is more useful than the Lebesgue integral. This integral was first defined by Arnaud Denjoy (1912). Denjoy was interested in a definition that would allow one to integrate functions like

$$f(x) = \frac{1}{x} \sin\left(\frac{1}{x^3}\right).$$

This function has a singularity at 0, and is not Lebesgue integrable. However, it seems natural to calculate its integral except over [-ε,δ] and then let ε, δ → 0.

Trying to create a general theory Denjoy used transfinite induction over the possible types of singularities which made the definition quite complicated. Other definitions were given by Nikolai Luzin (using variations on the notions of absolute continuity), and by Oskar Perron, who was interested in continuous major and minor functions. It took a while to understand that the Perron and Denjoy integrals are actually identical.

Later, in 1957, the Czech mathematician Jaroslav Kurzweil discovered a new definition of this integral elegantly similar in nature to Riemann's original definition which he named the gauge integral; the theory was developed by Ralph Henstock.

Due to these two important mathematicians, it is now commonly known as the Henstock–Kurzweil integral. The simplicity of Kurzweil's definition made some educators advocate that this integral should replace the Riemann integral in introductory calculus courses, but this idea has not gained traction.

Definition

Henstock's definition is as follows:

Given a tagged partition P of $[a, b]$, say

$$a = u_0 < u_1 < \cdots < u_n = b,\ t_i \in [u_{i-1}, u_i]$$

and a positive function

$$\delta : [a,b] \to (0,\infty),$$

which we call a *gauge*, we say P is δ-fine if

$$\forall i\ \ t_i - \delta(t_i) < u_{i-1} \le t_i \le u_i < t_i + \delta(t_i).$$

For a tagged partition P and a function

$$f : [a,b] \to \mathbb{R}$$

we define the Riemann sum to be

$$\sum_P f = \sum_{i=1}^{n} (u_i - u_{i-1}) f(t_i).$$

Given a function

$$f:[a,b]\rightarrow\mathbb{R},$$

we now define a number I to be the Henstock–Kurzweil integral of f if for every $\varepsilon > 0$ there exists a gauge δ such that whenever P is δ-fine, we have

$$\left|\sum_{P} f - I\right| < \varepsilon.$$

If such an I exists, we say that f is Henstock–Kurzweil integrable on $[a, b]$.

Cousin's theorem states that for every gauge δ, such a δ-fine partition P does exist, so this condition cannot be satisfied vacuously. The Riemann integral can be regarded as the special case where we only allow constant gauges.

Properties

Let f: $[a, b] \rightarrow$ R be any function.

If $a < c < b$, then f is Henstock–Kurzweil integrable on $[a, b]$ if and only if it is Henstock–Kurzweil integrable on both $[a, c]$ and $[c, b]$, and then

$$\int_a^b f(x)dx = \int_a^c f(x)dx + \int_c^b f(x)dx.$$

The Henstock–Kurzweil integral is linear, i.e., if f and g are integrable, and α, β are reals, then $\alpha f + \beta g$ is integrable and

$$\int_a^b \alpha f(x) + \beta g(x)dx = \alpha\int_a^b f(x)dx + \beta\int_a^b g(x)dx.$$

If f is Riemann or Lebesgue integrable, then it is also Henstock–Kurzweil integrable, and the values of the integrals are the same. The important Hake's theorem states that

$$\int_a^b f(x)dx = \lim_{c\rightarrow b-}\int_a^c f(x)dx$$

whenever either side of the equation exists, and symmetrically for the lower integration bound. This means that if f is "improperly Henstock–Kurzweil integrable", then it is properly Henstock–Kurzweil integrable; in particular, improper Riemann or Lebesgue integrals such as

$$\int_0^1 \frac{\sin(1/x)}{x} dx$$

are also Henstock–Kurzweil integrals. This shows that there is no sense in studying an "improper Henstock–Kurzweil integral" with finite bounds. However, it makes sense to consider improper Henstock–Kurzweil integrals with infinite bounds such as

$$\int_a^{+\infty} f(x)dx := \lim_{b \to +\infty} \int_a^b f(x)dx.$$

For many types of functions the Henstock–Kurzweil integral is no more general than Lebesgue integral. For example, if f is bounded, the following are equivalent:

- f is Henstock–Kurzweil integrable,
- f is Lebesgue integrable,
- f is Lebesgue measurable.

In general, every Henstock–Kurzweil integrable function is measurable, and f is Lebesgue integrable if and only if both f and $|f|$ are Henstock–Kurzweil integrable. This means that the Henstock–Kurzweil integral can be thought of as a "non-absolutely convergent version of Lebesgue integral".

It also implies that the Henstock–Kurzweil integral satisfies appropriate versions of the monotone convergence theorem (without requiring the functions to be nonnegative) and dominated convergence theorem (where the condition of dominance is loosened to $g(x) \leq f_n(x) \leq h(x)$ for some integrable g, h).

If F is differentiable everywhere (or with countable many exceptions), the derivative F2† is Henstock–Kurzweil integrable, and its indefinite Henstock–Kurzweil integral is F. (Note that F2† need not be Lebesgue integrable.)

In other words, we obtain a simpler and more satisfactory version of the second fundamental theorem of calculus: each differentiable function is, up to a constant, the integral of its derivative:

$$F(x) - F(a) = \int_a^x F'(t)dt.$$

Conversely, the Lebesgue differentiation theorem continues to holds for the Henstock–Kurzweil integral: if f is Henstock–Kurzweil integrable on $[a, b]$, and

$$F(x) = \int_a^x f(t)dt,$$

then $F'(x) = f(x)$ almost everywhere in $[a, b]$ (in particular, F is almost everywhere differentiable).

The space of all Henstock–Kurzweil-integrable functions is often endowed with the Alexiewicz norm, with respect to which it is barrelled but incomplete.

McShane Integral

Interestingly, Lebesgue integral on a line can also be presented in a similar fashion.

First of all, change of

$$\forall i \ u_i - u_{i-1} < \delta(t_i)$$

to

$$\forall i \ [u_{i-1}, u_i] \subset U_{\delta(t_i)}(t_i)$$

(here $U_\varepsilon(a)$ is a ε-neighbourhood of a) in the notion of δ-fine partition yields a definition of the Henstock–Kurzweil integral equivalent to the one given above. But after this change we can drop condition

$$t_i \in [u_{i-1}, u_i]$$

and get a definition of *McShane integral*, which is equivalent to the Lebesgue integral.

Chapter 6

Quantum Calculus

Quantum calculus, sometimes called calculus without limits, is equivalent to traditional infinitesimal calculus without the notion of limits. It defines "q-calculus" and "h-calculus". h ostensibly stands for Planck's constant while q stands for quantum. The two parameters are related by the formula

$$q = e^{ih} = e^{2\pi i\hbar}$$

where $\hbar = \dfrac{h}{2\pi}$ is the reduced Planck constant.

Differentiation

In the q-calculus and h-calculus, differentials of functions are defined as

$$d_h(f(x)) = f(x+h) - f(x)$$

and

$$d_h(f(x)) = f(x+h) - f(x)$$

respectively. Derivatives of functions are then defined as fractions by the q-derivative

$$D_q(f(x)) = \frac{d_q(f(x))}{d_q(x)} = \frac{f(qx) - f(x)}{(q-1)x}$$

and by

$$D_h(f(x)) = \frac{d_h(f(x))}{d_h(x)} = \frac{f(x+h) - f(x)}{h}$$

In the limit, as h goes to 0, or equivalently as q goes to 1, these expressions take on the form of the derivative of classical calculus.

Integration

q-integral: A function F(x) is a q-antiderivative of f(x) if $D_q F(x)=f(x)$. The q-antiderivative (or q-integral) is denoted by $\int f(x)d_q x$ and an expression for F(x) can be found from the formula $\int f(x)d_q x=(1-q)\sum_{j=0}^{\infty} xq^j f(xq^j)$ which is called the Jackson integral of f(x). For 0 < q < 1, the series converges to a function F(x) on an interval (0,A] if |f(x)x^α| is bounded on the interval (0,A] for some 0 <= α < 1.

The q-integral is a Riemann-Stieltjes integral with respect to a step function having infinitely many points of increase at the points q^j, with the jump at the point q^j being q^j. If we call this step function $g_q(t)$ then $dg_q(t) = d_q t$.

h-integral

A function F(x) is an h-antiderivative of f(x) if $D_h F(x)=f(x)$. The h-antiderivative (or h-integral) is denoted by $\int f(x)d_h x$. If a and b differ by an integer multiple of h then the definite integral $\int_a^b f(x)d_h x$ is given by a Riemann sum of f(x) on the interval [a,b] partitioned into subintervals of width h.

Example

The derivative of the function x^n (for some positive integer *n*) in the classical calculus is nx^{n-1}. The corresponding expressions in q-calculus and h-calculus are

$$D_q(x^n)=\frac{q^n-1}{q-1}x^{n-1}=[n]_q\,x^{n-1}$$

with the q-bracket

$$[n]_q = \frac{q^n - 1}{q - 1}$$

and

$$D_h(x^n) = x^{n-1} + hx^{n-2} + \cdots + h^{n-1}$$

respectively. The expression $[n]_q x^{n-1}$ is then the q-calculus analogue of the simple power rule for positive integral powers. In this sense, the function x^n is still *nice* in the q-calculus, but rather ugly in the h-calculus – the h-calculus analog of x^n is instead the falling factorial $(x)_n := x(x-1)\cdots(x-n+1).$,

One may proceed further and develop, for example, equivalent notions of Taylor expansion, et cetera, and even arrive at q-calculus analogues for all of the usual functions one would want to have, such as an analogue for the sine function whose q-derivative is the appropriate analogue for the cosine.

History

The h-calculus is just the calculus of finite differences, which had been studied by George Boole and others, and has proven useful in a number of fields, among them combinatorics and fluid mechanics. The q-calculus, while dating in a sense back to Leonhard Euler and Carl Gustav Jacobi, is only recently beginning to see more usefulness in quantum mechanics, having an intimate connection with commutativity relations and Lie algebra.

Differentiation Rules

This is a summary of differentiation rules, that is, rules for computing the derivative of a function in calculus.

Elementary Rules of Differentiation

Unless otherwise stated, all functions are functions of real numbers (R) that return real values; although more generally, the formulae below apply wherever they are well defined—including complex numbers (C).

Differentiation is Linear

For any functions f and g and any real numbers a and b the derivative of the function $h(x) = af(x) + bg(x)$ with respect to x is

$$h'(x) = af'(x) + bg'(x).$$

In Leibniz's notation this is written as:

$$\frac{d(af + bg)}{dx} = a\frac{df}{dx} + b\frac{dg}{dx}.$$

Special cases include:

- *The constant multiple rule*

$$(af)' = af'$$

- *The sum rule*

$$(f + g)' = f' + g'$$

- *The subtraction rule*

$$(f - g)' = f' - g'.$$

The Product Rule (Leibniz Rule)

For the functions f and g, the derivative of the function $h(x) = f(x)\, g(x)$ with respect to x is

$$h'(x) = f'(x)g(x) + f(x)g'(x).$$

In Leibniz's notation this is written

$$\frac{d(fg)}{dx} = \frac{df}{dx}g + f\frac{dg}{dx}.$$

The Chain Rule

The derivative of the function of a function $h(x) = f(g(x))$ with respect to x is

$$h'(x) = f'(g(x))g'(x).$$

In Leibniz's notation this is written as:

$$\frac{dh}{dx} = \frac{df}{dg}\frac{dg}{dx}.$$

However, by relaxing the interpretation of h as a function, this is often simply written

$$\frac{dh}{dx} = \frac{dh}{dg}\frac{dg}{dx}.$$

The Inverse Function Rule

If the function f has an inverse function g, meaning that $g(f(x)) = x$ and $f(g(y)) = y$, then

$$g' = \frac{1}{f' \circ g}.$$

In Leibniz notation, this is written as

$$\frac{dx}{dy} = \frac{1}{dy / dx}.$$

Power laws, polynomials, quotients, and reciprocals

The Polynomial or Elementary Power Rule

If $f(x) = x^n$, for any integer n then

$$f'(x) = nx^{n-1}.$$

Special cases include:

- *Constant rule*: if f is the constant function $f(x) = c$, for any number c, then for all x, $f2\ (x) = 0$.
- if $f(x) = x$, then $f2\ (x) = 1$. This special case may be generalized to:

The derivative of a linear function is constant: if $f(x) = ax + b$, then $f2\ (x) = a$.

Combining this rule with the linearity of the derivative permits the computation of the derivative of any polynomial.

The Reciprocal Rule

The derivative of $h(x) = 1/f(x)$ for any (nonvanishing) function f is:

$$h'(x) = -\frac{f'(x)}{[f(x)]^2}.$$

In Leibniz's notation, this is written

$$\frac{d(1/f)}{dx} = -\frac{1}{f^2}\frac{df}{dx}.$$

The reciprocal rule can be derived from the chain rule and the power rule.

The Quotient Rule

If f and g are functions, then:

$$\left(\frac{f}{g}\right)' = \frac{f'g - g'f}{g^2} \quad \text{wherever } g \text{ is nonzero.}$$

This can be derived from reciprocal rule and the product rule. Conversely (using the constant rule) the reciprocal rule may be derived from the special case $f(x) = 1$.

Generalized Power Rule

The elementary power rule generalizes considerably. The most general power rule is the functional power rule: for any functions f and g,

$$(f^g)' = \left(e^{g \ln f}\right)' = f^g\left(f'\frac{g}{f} + g' \ln f\right),$$

wherever both sides are well defined.

Special cases:

- If $f(x) = x^a$, $f2\,(x) = ax^{a-1}$ when a is any real number and x is positive.
- The reciprocal rule may be derived as the special case where $g(x) = -1$.

Derivatives of Exponential and Logarithmic Functions

$$\left(c^{ax}\right)' = c^{ax} \ln c \cdot a, \qquad c > 0$$

note that the equation above is true for all c, but the derivative for c < 0 yields a complex number.

$$\left(e^x\right)' = e^x$$

$$\left(\log_c x\right)' = \frac{1}{x \ln c}, \qquad c > 0, c \neq 1$$

the equation above is also true for all c but yields a complex number.

$$\left(\ln x\right)' = \frac{1}{x}, \qquad x \neq 0$$

$$(\ln|x|)' = \frac{1}{x}$$

$$\left(x^x\right)' = x^x(1+\ln x)$$

The derivative of the natural logarithm with a generalised functional argument *f*(*x*) is

$$\frac{d}{dx}[\ln(f(x))] = \frac{f'(x)}{f(x)}$$

By applying the change-of-base identity, the derivative for other bases is

$$\frac{d}{dx}\log_b(x) = \frac{d}{dx}\frac{\ln(x)}{\ln(b)} = \frac{1}{x\ln(b)} = \frac{\log_b(e)}{x}.$$

Logarithmic Derivatives

The logarithmic derivative is another way of stating the rule for differentiating the logarithm of a function (using the chain rule):

$$(\ln f)' = \frac{f'}{f} \quad \text{wherever } f \text{ is positive.}$$

Derivatives to nth Order

Some rules exist for computing the *n*th derivative of functions, where *n* is a positive integer. These include:

Faà di Bruno's Formula

If *f* and *g* are *n* times differentiable, then

$$\frac{d^n}{dx^n}[f(g(x))] = n!\sum_{\{k_m\}} f^{(r)}(g(x))\prod_{m=1}^{n}\frac{1}{k_m!}\left(g^{(m)}(x)\right)^{k_m}$$

where $r = \sum_{m=1}^{n-1} k_m$ and the set $\{k_m\}$ consists of all non-negative integer solutions of the Diophantine equation $\sum_{m=1}^{n} mk_m = n$.

General Leibniz Rule

If f and g are n times differentiable, then

$$\frac{d^n}{dx^n}[f(x)g(x)] = \sum_{k=0}^{n} \binom{n}{k} \frac{d^{n-k}}{dx^{n-k}} f(x) \frac{d^k}{dx^k} g(x)$$

Benjamin–Bona–Mahony Equation

The Benjamin–Bona–Mahony equation (or BBM equation) is the partial differential equation

$$u_t + u_x + uu_x - u_{xxt} = 0.$$

This equation was introduced in Benjamin, Bona, and Mahony (1972) as an improvement of the Korteweg–de Vries equation (KdV equation) for modeling long waves of small amplitude in 1+1 dimensions. They show the stability and uniqueness of solutions to the BBM equation. This contrasts with the KdV equation, which is unstable in its high wavenumber components. Further, while the KdV equation has an infinite number of integrals of motion, the BBM equation only has three.

A higher-dimensional version is given by

$$u_t - \nabla^2 u_t + \mathrm{div}\varphi(u) = 0.$$

where $\varphi$is a fixed smooth function from Rto R^n. Avrin & Goldstein (1985) proved local existence of a solution in all dimensions.

Related Rates

In differential calculus, related rates problems involve finding a rate at which a quantity changes by relating that quantity to other quantities whose rates of change are known. The rate of change is usually with respect to time. Because science and engineering often relate quantities to each other, the methods of related rates have broad applications in these fields. Because problems to which related rate methods apply often have nested functions, solving these problems often requires application of the chain rule.

Procedure

The most common way to approach related rates problems is the following:

1. Identify the known variables, including rates of change and the rate of change that is to be found. (Drawing a picture or representation of the problem can help to keep everything in order)
2. Construct an equation relating the quantities whose rates of change are known to the quantity whose rate of change is to be found.
3. Differentiate both sides of the equation with respect to time (or other rate of change). Often, the chain rule is employed at this step.
4. Substitute the known rates of change and the known quantities into the equation.
5. Solve for the wanted rate of change.

Errors in this procedure are often caused by plugging in the known values for the variables *before* (rather than after) finding the derivative with respect to time. Doing so will yield an incorrect result, since if those values are substituted for the variables before differentiation, those variables will become constants; and when the equation is differentiated, zeroes appear in places of all variables for which the values were plugged in.

Examples

Leaning Ladder Example: A 10-meter ladder is leaning against the wall of a building, and the base of the ladder is sliding away from the building at a rate of 3 meters per second. How fast is the top of the ladder sliding down the wall when the base of the ladder is 6 meters from the wall?

The distance between the base of the ladder and the wall, x, and the height of the ladder on the wall, y, represent the sides of a right triangle with the ladder as the hypotenuse, h. The object is to find dy/dt, the rate of change of y with respect to time, t, when h, x and dx/dt, the rate of change of x, are known.

Step 1:

$$x = 6$$

$$h = 10$$

$$\frac{dx}{dt} = 3$$

$$\frac{dh}{dt} = 0$$

$$\frac{dy}{dt} = ?$$

Step 2: From the Pythagorean theorem, the equation

$$x^2 + y^2 = h^2$$

describes the relationship between x, y and h, for a right triangle. Differentiating both sides of this equation with respect to time, t, yields

$$\frac{d}{dt}(x^2 + y^2) = \frac{d}{dt}(h^2)$$

Step 3: When solved for the wanted rate of change, dy/dt, gives us

$$\frac{d}{dt}(x^2) + \frac{d}{dt}(y^2) = \frac{d}{dt}(h^2)$$

$$(2x)\frac{dx}{dt} + (2y)\frac{dy}{dt} = (2h)\frac{dh}{dt}$$

$$x\frac{dx}{dt} + y\frac{dy}{dt} = h\frac{dh}{dt}$$

$$\frac{dy}{dt} = \frac{h\frac{dh}{dt} - x\frac{dx}{dt}}{y}.$$

Step 4 & 5: Using the variables from step 1 gives us:

$$\frac{dy}{dt} = \frac{h\frac{dh}{dt} - x\frac{dx}{dt}}{y}.$$

$$\frac{dy}{dt} = \frac{10 \times 0 - 6 \times 3}{y} = -\frac{18}{y}.$$

Solving for y using the Pythagorean Theorem gives:

$$x^2 + y^2 = h^2$$

$$6^2 + y^2 = 10^2$$

$$y = 8$$

Plugging in 8 for the equation:

$$-\frac{18}{y} = -\frac{18}{8} = -\frac{9}{4}$$

It is generally assumed that negative values represent the downward direction. In doing such, the top of the ladder is sliding down the wall at a rate of ^{9}D $_{4}$ meters per second.

Physics Examples

Because one physical quantity often depends on another, which, in turn depends on others, such as time, related rate methods have broad applications in Physics. This section presents an example of related rates kinematics and electromagnetic induction.

Physics Example II: Electromagnetic Induction of Conducting Loop Spinning in Magnetic Field

The magnetic flux through a loop of area A whose normal is at an angle θ to a magnetic field of strength B is

$$\Phi_B = BA\cos(\theta),$$

Faraday's law of electromagnetic induction states that the induced electromotive force ε is the negative rate of change of magnetic flux Φ_B through a conducting loop.

$$\mathcal{E} = -\frac{d\Phi_B}{dt},$$

If the loop area A and magnetic field B are held constant, but the loop is rotated so that the angle θ is a known function of time, the rate of change of θ can be related to the rate of change of Φ_B (and therefore the electromotive force) by taking the time derivative of the flux relation

$$\mathcal{E} = -\frac{d\Phi_B}{dt} = BA\sin\theta\frac{d\theta}{dt}$$

If for example, the loop is rotating at a constant angular velocity Ω, so that θ=ωt, then

$$\mathcal{E} = \omega BA\sin\omega t$$

Differential of a Function

In calculus, the differential represents the principal part of the change in a function $y = f(x)$ with respect to changes in the independent variable. The differential dy is defined by

$$dy = f'(x)dx,$$

where $f'(x)$ is the derivative of f with respect to x, and dx is an additional real variable (so that dy is a function of x and dx). The notation is such that the equation

$$dy = \frac{dy}{dx}dx$$

holds, where the derivative is represented in the Leibniz notation dy/dx, and this is consistent with regarding the derivative as the quotient of the differentials. One also writes

$$df(x) = f'(x)dx.$$

The precise meaning of the variables dy and dx depends on the context of the application and the required level of mathematical rigor.

The domain of these variables may take on a particular geometrical significance if the differential is regarded as a particular differential form, or analytical significance if the differential is regarded as a linear approximation to the increment of a function. In physical applications, the variables dx and dy are often constrained to be very small ("infinitesimal").

History and Usage

The differential was first introduced via an intuitive or heuristic definition by Gottfried Wilhelm Leibniz, who thought of the differential dy as an infinitely small (or infinitesimal) change in the value y of the function, corresponding to an infinitely small change dx in the function's argument x. For that reason, the instantaneous rate of change of y with respect to x, which is the value of the derivative of the function, is denoted by the fraction

$$\frac{dy}{dx}$$

in what is called the Leibniz notation for derivatives. The quotient dy/dx is not infinitely small; rather it is a real number.

The use of infinitesimals in this form was widely criticized, for instance by the famous pamphlet The Analyst by Bishop Berkeley. Augustin-Louis Cauchy (1823) defined the differential without appeal to the atomism of Leibniz's infinitesimals. Instead, Cauchy, following d'Alembert, inverted the logical order of Leibniz and his successors: the derivative itself became the fundamental object, defined as a limit of difference quotients, and the differentials were then defined in terms of it. That is, one was free to *define* the differential dy by an expression

$$dy = f'(x)dx$$

in which dy and dx are simply new variables taking finite real values, not fixed infinitesimals as they had been for Leibniz.

According to Boyer (1959, p. 12), Cauchy's approach was a significant logical improvement over the infinitesimal approach of Leibniz because, instead of invoking the metaphysical notion of infinitesimals, the quantities dy and dx could now be manipulated in exactly the same manner as any other real quantities in a meaningful way.

Cauchy's overall conceptual approach to differentials remains the standard one in modern analytical treatments, although the final word on rigor, a fully modern notion of the limit, was ultimately due to Karl Weierstrass.

In physical treatments, such as those applied to the theory of thermodynamics, the infinitesimal view still prevails. Courant & John (1999, p. 184) reconcile the physical use of infinitesimal differentials with the mathematical impossibility of them as follows. The differentials represent finite non-zero values that are smaller than the degree of accuracy required for the particular purpose for which they are intended. Thus "physical infinitesimals" need not appeal to a corresponding mathematical infinitesimal in order to have a precise sense. Following twentieth-century developments in mathematical analysis and differential geometry, it became clear that the notion of the differential of a function could be extended in a variety of ways. In real analysis, it is more desirable to deal directly with the differential as the principal part of the increment of a function. This leads directly to the notion that the differential of a function at a point is linear functional of an increment Δx.

This approach allows the differential (as a linear map) to be developed for a variety of more sophisticated spaces, ultimately giving rise to such notions as the Fréchet or Gâteaux derivative.

Likewise, in differential geometry, the differential of a function at a point is a linear function of a tangent vector (an "infinitely small displacement"), which exhibits it as a kind of one-form: the exterior derivative of the function.

In non-standard calculus, differentials are regarded as infinitesimals, which can themselves be put on a rigorous footing.

Definition

The differential is defined in modern treatments of differential calculus as follows. The differential of a function $f(x)$ of a single real variable x is the function df of two independent real variables x and Δx given by

$$df(x,\Delta x) \stackrel{\text{def}}{=} f'(x)\Delta x.$$

One or both of the arguments may be suppressed, i.e., one may see $df(x)$ or simply df. If $y = f(x)$, the differential may also be written as dy. Since $dx(x, \Delta x) = \Delta x$ it is conventional to write $dx = \Delta x$, so that the following equality holds:

$$df(x) = f'(x)dx$$

This notion of differential is broadly applicable when a linear approximation to a function is sought, in which the value of the increment Δx is small enough. More precisely, if f is a differentiable function at x, then the difference in y-values

$$\Delta y \stackrel{\text{def}}{=} f(x+\Delta x) - f(x)$$

satisfies

$$\Delta y = f'(x)\Delta x + \varepsilon = df(x) + \varepsilon$$

where the error ε in the approximation satisfies $\varepsilon/\Delta x \to 0$ as $\Delta x \to 0$. In other words, one has the approximate identity

$$\Delta y \approx dy$$

in which the error can be made as small as desired relative to Δx by constraining Δx to be sufficiently small; that is to say,

$$\frac{\Delta y - dy}{\Delta x} \to 0$$

as $\Delta x \to 0$. For this reason, the differential of a function is known as the principal (linear) part in the increment of a function: the differential is a linear function of the increment Δx, and although the error ε may be nonlinear, it tends to zero rapidly as Δx tends to zero.

Differentials in Several Variables

Following Goursat (1904, I, §15), for functions of more than one independent variable,

$$y = f(x_1, \ldots, x_n),$$

the partial differential of y with respect to any one of the variables x_1 is the principal part of the change in y resulting from a change dx_1 in that one variable. The partial differential is therefore

$$\frac{\partial y}{\partial x_1} dx_1$$

involving the partial derivative of y with respect to x_1. The sum of the partial differentials with respect to all of the independent variables is the total differential

$$dy = \frac{\partial y}{\partial x_1} dx_1 + \cdots + \frac{\partial y}{\partial x_n} dx_n,$$

which is the principal part of the change in y resulting from changes in the independent variables x_i.

More precisely, in the context of multivariable calculus, following Courant (1937ii), if f is a differentiable function, then by the definition of the differentiability, the increment

$$\begin{aligned}\Delta y &\overset{\text{def}}{=} f(x_1+\Delta x_1,\ldots,x_n+\Delta x_n)-f(x_1,\ldots,x_n)\\ &=\frac{\partial y}{\partial x_1}\Delta x_1+\cdots+\frac{\partial y}{\partial x_n}\Delta x_n+\varepsilon_1\Delta x_1+\cdots+\varepsilon_n\Delta x_n\end{aligned}$$

where the error terms ε_i tend to zero as the increments Δx_i jointly tend to zero. The total differential is then rigorously defined as

$$dy=\frac{\partial y}{\partial x_1}\Delta x_1+\cdots+\frac{\partial y}{\partial x_n}\Delta x_n.$$

Since, with this definition,

$$dx_i(\Delta x_1,\ldots,\Delta x_n)=\Delta x_i,$$

one has

$$dy=\frac{\partial y}{\partial x_1}dx_1+\cdots+\frac{\partial y}{\partial x_n}dx_n.$$

As in the case of one variable, the approximate identity holds

$$dy\approx\Delta y$$

in which the total error can be made as small as desired relative to $\sqrt{\Delta x_1^2+\cdots+\Delta x_n^2}$ by confining attention to sufficiently small increments.

Higher-order Differentials

Higher-order differentials of a function $y = f(x)$ of a single variable x can be defined via:

$$d^2y=d(dy)=d(f'(x)dx)=f''(x)(dx)^2,$$

and, in general,

$$d^ny=f^{(n)}(x)(dx)^n.$$

Informally, this justifies Leibniz's notation for higher-order derivatives

$$f^{(n)}(x)=\frac{d^nf}{dx^n}.$$

When the independent variable x itself is permitted to depend on other variables, then the expression becomes more complicated, as it must include also higher order differentials in x itself. Thus, for instance,

$$\begin{aligned} d^2y &= f''(x)(dx)^2 + f'(x)d^2x \\ d^3y &= f'''(x)(dx)^3 + 3f''(x)dxd^2x + f'(x)d^3x \end{aligned}$$

and so forth.

Similar considerations apply to defining higher order differentials of functions of several variables. For example, if f is a function of two variables x and y, then

$$d^n f = \sum_{k=0}^{n} \binom{n}{k} \frac{\partial^k f}{\partial x^k} \frac{\partial^{n-k} f}{\partial y^{n-k}} (dx)^k (dy)^{n-k},$$

where $\binom{n}{k}$ is a binomial coefficient. In more variables, an analogous expression holds, but with an appropriate multinomial expansion rather than binomial expansion.

Higher order differentials in several variables also become more complicated when the independent variables are themselves allowed to depend on other variables. For instance, for a function f of x and y which are allowed to depend on auxiliary variables, one has

$$d^2 f = \left(\frac{\partial^2 f}{\partial x^2}(dx)^2 + 2\frac{\partial^2 f}{\partial x \partial y} dxdy + \frac{\partial^2 f}{\partial y^2}(dy)^2 \right) + \frac{\partial f}{\partial x} d^2x + \frac{\partial f}{\partial y} d^2y.$$

Because of this notational infelicity, the use of higher order differentials was roundly criticized by Hadamard 1935, who concluded:

Enfin, que signifie ou que représente l'égalité

$$d^2z = r\,dx^2 + 2s\,dxdy + t\,dy^2?$$

A mon avis, rien du tout.

That is: *Finally, what is meant, or represented, by the equality [...]? In my opinion, nothing at all.* In spite of this skepticism, higher order differentials did emerge as an important tool in analysis

In these contexts, the nth order differential of the function f applied to an increment Δx is defined by

$$d^n f(x, \Delta x) = \frac{d^n}{dt^n} f(x + t\Delta x)\bigg|_{t=0}$$

or an equivalent expression, such as

$$\lim_{t \to 0} \frac{\Delta^n_{t\Delta x} f}{t^n}$$

where $\Delta^n_{t\Delta x} f$ is an nth forward difference with increment $t\Delta x$.

This definition makes sense as well if f is a function of several variables (for simplicity taken here as a vector argument). Then the nth differential defined in this way is a homogeneous function of degree n in the vector increment Δx. Furthermore, the Taylor series of f at the point x is given by

$$f(x + \Delta x) \sim f(x) + df(x, \Delta x) + \frac{1}{2} d^2 f(x, \Delta x) + \cdots + \frac{1}{n!} d^n f(x, \Delta x) + \cdots$$

The higher order Gâteaux derivative generalizes these considerations to infinite dimensional spaces.

Properties

A number of properties of the differential follow in a straightforward manner from the corresponding properties of the derivative, partial derivative, and total derivative. These include:

Linearity: For constants a and b and differentiable functions f and g,

$$d(af + bg) = a\,df + b\,dg.$$

Product Rule: For two differentiable functions f and g,

$$d(fg) = f\,dg + g\,df.$$

An operation d with these two properties is known in abstract algebra as a derivation. In addition, various forms of the chain rule hold, in increasing level of generality:

If $y = f(u)$ is a differentiable function of the variable u and $u = g(x)$ is a differentiable function of x, then

$$dy = f'(u)du = f'(g(x))g'(x)dx.$$

If $y = f(x_1, ..., x_n)$ and all of the variables $x_1, ..., x_n$ depend on another variable t, then by the chain rule for partial derivatives, one has

$$\begin{aligned} dy &= \frac{dy}{dt} dt \\ &= \frac{\partial y}{\partial x_1} dx_1 + \cdots + \frac{\partial y}{\partial x_n} dx_n \\ &= \frac{\partial y}{\partial x_1} \frac{dx_1}{dt} dt + \cdots + \frac{\partial y}{\partial x_n} \frac{dx_n}{dt} dt. \end{aligned}$$

Heuristically, the chain rule for several variables can itself be understood by dividing through both sides of this equation by the infinitely small quantity dt. More general analogous expressions hold, in which the intermediate variables x_i depend on more than one variable.

General Formulation

A consistent notion of differential can be developed for a function f R^n → R^m between two Euclidean spaces. Let x,Δx ∈ R^n be a pair of Euclidean vectors. The increment in the function f is

$$\Delta f = f(\mathbf{x} + \Delta \mathbf{x}) - f(\mathbf{x}).$$

If there exists an $m \times n$ matrix A such that

$$\Delta f = A\Delta\mathbf{x} + \|\Delta\mathbf{x}\|\,\boldsymbol{\varepsilon}$$

in which the vector e ® 0 as Dx ® 0, then f is by definition differentiable at the point x. The matrix A is sometimes known as the Jacobian matrix, and the linear transformation that associates to the increment Dx ∈ R^n the vector ADx ∈ R^m is, in this general setting, known as the differential $df(x)$ of f at the point x. This is precisely the Fréchet derivative, and the same construction can be made to work for a function between any Banach spaces.

Another fruitful point of view is to define the differential directly as a kind of directional derivative:

$$df(\mathbf{x},\mathbf{h}) = \lim_{t\to 0} \frac{f(\mathbf{x}+t\mathbf{h}) - f(\mathbf{x})}{t} = \frac{d}{dt} f(\mathbf{x}+t\mathbf{h})\Big|_{t=0},$$

which is the approach already taken for defining higher order differentials (and is most nearly the definition set forth by Cauchy). If t represents time and x position, then h represents a velocity instead of a displacement as we have heretofore regarded it. This yields yet another refinement of the notion of differential: that it should be a linear function of a kinematic velocity. The set of all velocities through a given point of space is known as the tangent space, and so df gives a linear function on the tangent space: a differential form. With this interpretation, the differential of f is known as the exterior derivative, and has broad application in differential geometry because the notion of velocities and the tangent space makes sense on any differentiable manifold. If, in addition, the output value of f also represents a position (in a Euclidean space), then a dimensional analysis confirms that the output value of df must be a velocity. If one treats the differential in this manner, then it is known as the pushforward since it "pushes" velocities from a source space into velocities in a target space.

Other Approaches

Although the notion of having an infinitesimal increment dx is not well-defined in modern mathematical analysis, a variety of techniques exist for defining the infinitesimal differential so that the differential of a function can be handled in a manner that does not clash with the Leibniz notation. These include: Defining the differential as a kind of differential form, specifically the exterior derivative of a function. The infinitesimal increments are then identified with vectors in the tangent space at a point. This approach is popular in differential geometry and related fields, because it readily generalizes to mappings between differentiable manifolds.

Differentials as nilpotent elements of commutative rings. This approach is popular in algebraic geometry. Differentials in smooth models of set theory. This approach is known as synthetic differential geometry or smooth infinitesimal analysis and is closely related to the algebraic geometric approach, except that ideas from topos theory are used to *hide* the mechanisms by which nilpotent infinitesimals are introduced.

Differentials as infinitesimals in hyperreal number systems, which are extensions of the real numbers which contain invertible infinitesimals and infinitely large numbers. This is the approach of nonstandard analysis pioneered by Abraham Robinson.

Examples and Applications

Differentials may be effectively used in numerical analysis to study the propagation of experimental errors in a calculation, and thus the overall numerical stability of a problem (Courant 1937i). Suppose that the variable x represents the outcome of an experiment and y is the result of a numerical computation applied to x. The question is to what extent errors in the measurement of x influence the outcome of the computation of y. If the x is known to within Δx of its true value, then Taylor's theorem gives the following estimate on the error Δy in the computation of y:

$$\Delta y = f'(x)\Delta x + \frac{(\Delta x)^2}{2} f''(\xi)$$

where $\partial = x + \theta\Delta x$ for some $0 < \theta < 1$. If Δx is small, then the second order term is negligible, so that Δy is, for practical purposes, well-approximated by $dy = f'(x)\Delta x$.

The differential is often useful to rewrite a differential equation

$$\frac{dy}{dx} = g(x)$$

in the form

$$dy = g(x)dx,$$

in particular when one wants to separate the variables.

Implicit and Explicit Functions

The implicit function theorem provides a link between implicit and explicit functions. It states that if the equation $R(x, y) = 0$ satisfies some mild conditions on its partial derivatives, then one can in principle solve this equation for y, at least over some small interval. Geometrically, the graph defined by $R(x,y) = 0$ will overlap locally with the graph of an equation $y = f(x)$.

Various numerical methods exist for solving the equation $R(x,y)=0$ to find an approximation to the implicit function f.

Many of these methods are iterative in that they produce successively better approximations, so that a prescribed accuracy can be achieved. Many of these iterative methods are based on some form of Newton's method.

Examples

Inverse Functions: A common type of implicit function is an inverse function. If f is a function, then the inverse function of f, called f^{-1}, is the function giving a solution of the equation

$$x = f(y)$$

for y in terms of x. This solution is

$$y = f^{-1}(x).$$

Intuitively, an inverse function is obtained from f by interchanging the roles of the dependent and independent variables. Stated another way, the inverse function gives the solution for y of the equation

$$R(x, y) = x - f(y) = 0.$$

Examples

1. The natural logarithm $\ln(x)$ gives the solution $y = \ln(x)$ of the equation $x - e^y = 0$ or equivalently of $x = e^y$. Here

   $f(y) = e^y$ and $f^{-1}(x) = \ln(x)$.

2. The product log is an implicit function giving the solution for y of the equation $x - y\ e^y = 0$.

Algebraic Functions

An algebraic function is a function that satisfies a polynomial equation whose coefficients are themselves polynomials. For example, an algebraic function in one variable x gives a solution for y of an equation

$$a_n(x)y^n + a_{n-1}(x)y^{n-1} + \cdots + a_0(x) = 0$$

where the coefficients $a_i(x)$ are polynomial functions of x. Algebraic functions play an important role in mathematical analysis and algebraic geometry. A simple example of an algebraic function is given by the unit circle equation:

$$x^2 + y^2 - 1 = 0.$$

Solving for y gives an explicit solution:

$$y = \pm\sqrt{1 - x^2}.$$

While explicit solutions can be found for equations that are quadratic, cubic, and quartic in y, the same is not in general true for quintic and higher degree equations, such as

$$y^5 + 2y^4 - 7y^3 + 3y^2 - 6y - x = 0.$$

Nevertheless, one can still refer to the implicit solution $y = g(x)$ involving the multi-valued implicit function g.

Caveats

Not every equation $R(x, y) = 0$ implies a graph of a single-valued function, the circle equation being one prominent example. Another example is an implicit function given by $x - C(y) = 0$ where C is a cubic polynomial having a "hump" in its graph. Thus, for an implicit function to be a *true* (single-valued) function it might be necessary to use just part of the graph. An implicit function can sometimes be successfully defined as a true function only after "zooming in" on some part of the x-axis and "cutting away" some unwanted function branches. Then an equation expressing y as an implicit function of the other variable(s) can be written.

The defining equation $R(x, y) = 0$ can also have other pathologies. For example, the equation $x = 0$ does not imply a function $f(x)$ giving solutions for y at all; it is a vertical line. In order to avoid a problem like this, various constraints are frequently imposed on the allowable sorts of equations or on the domain. The implicit function theorem provides a uniform way of handling these sorts of pathologies.

Chapter 7

Convergent Series

Definition: Let $\{a_n\}$ be a sequence, and let,

$S_n = a_1 + a_2 + \cdots + a_n$, the nth partial sum.

If $\lim_{n \to \infty} s_n$ exists, we say that $\sum a_n$ is a convergent series,

and write $\lim_{n \to \infty} s_n = \sum a_n$.

Thus a series is convergent if and only if it's sequence of partial sums is convergent. The limit of the sequence of partial sums is the sum of the series. A series which is *not* convergent, is a divergent series.

Example: The series $\sum r^n$ is convergent with sum $1/(1-r)$, provided that $|r| < 1$. For other values of r, the series is divergent; in particular, the series $\sum(-1)^n$ is divergent.

Solution: We noted above that when $|r| < 1$, $S_n \to a/(1-r)$ as $n \to \infty$; note particular cases;

$$\sum_{n=1}^{\infty} \frac{1}{2^n} = 1 \text{ or equivalently, } \frac{1}{2} + \frac{1}{4} + \frac{1}{8} + \ldots = 1.$$

Example: The sum $\sum \frac{1}{n(n+1)}$ is convergent with sum

1.

Solution: We can compute the partial sums explicitly:

$$S_n = \sum_{k=1}^{n} \frac{1}{k(k+1)} = \sum_{k=1}^{n} \left(\frac{1}{k} - \frac{1}{k+1} \right) = 1 - \frac{1}{n+1} \to 1 \text{ as } n \to \infty$$

Example: The sum $\sum \frac{1}{n}$ is divergent.

Solution: We estimate the partial sums:

$$S_n = \frac{1}{1} + \frac{1}{2} + \left(\frac{1}{3} + \frac{1}{4} \right) + \left(\frac{1}{5} + \ldots + \frac{1}{8} \right) + \left(\frac{1}{9} + \ldots + \frac{1}{15} \right) + \ldots + \frac{1}{n}$$

$$> 1 + \frac{1}{2} + \frac{2}{4} + \frac{4}{8} > 2\frac{1}{2} \text{ if } n \geq 15$$

$$> 1 + \frac{1}{2} + \frac{2}{4} + \frac{4}{8} + \frac{8}{16} > 3$$

if $n \geq 31$

$$\to \infty \text{ as } n \to \infty .$$

Example: The sum is $\sum \frac{1}{n^2}$ convergent. [Actually the sum is $\frac{\pi^2}{6}$, but this is much harder.]

Solution: We estimate the partial sums. Since $1/n^2 > 0$, clearly $\{S_n\}$ is an increasing sequence. We show it is bounded above, whence by the Monotone Convergence Theorem, it is convergent. From the diagram,

$$\frac{1}{2^2} + \frac{1}{3^2} + \ldots + \frac{1}{n^2} = \int_1^n \frac{dx}{x^2} \text{, and so}$$

$$S_n < 1 + \left[-\frac{1}{x} \right]_1^n \leq 2 - \frac{1}{n}$$

Thus $S_n < 2$ for all n, the sequence of partial sums is bounded above, and the series is convergent.

Proposition: Let $\sum a_n$ be convergent. Then $a_n \to 0$ as $n \to \infty$.

Proof: Write $l = \lim_{n\to\infty} s_n$, and recall from our work on limits of sequences that $S_{n-1} l \to l$ as $n \to \infty$.

Then,

$$\begin{aligned} a_n &= (a_1 + a_2 + \ldots a_n) - (a_1 + a_2 + \ldots a_{n-1}) \\ &= S_n - S_{n-1} l - l \quad \text{as} \quad n. \end{aligned}$$

Example: The sum $\sum \frac{1}{n}$ is divergent (Graphical method).

Solution: We estimate the partial sums. Since $1/n > 0$, clearly $\{S_n\}$ is an increasing sequence. We show it is not bounded above, whence by the note after 3.9, the sequence of partial sums $\to \infty$ as $n \to \infty$.

From the diagram,

$$1 + \frac{1}{2} + \ldots + \frac{1}{n} > \int_1^n \frac{dx}{x} > \frac{1}{2} + \ldots + \frac{1}{n}$$

Writing $S_n = 1 + \frac{1}{2} + \cdots + \frac{1}{n}$, we have $S_n > \log n > S_n - 1$, or equivalently $1 + \log n > S_n > \log n$ for all n. Thus $S_n \to \infty$ and the series is divergent.

Proposition: Let $\sum a_n$ and $\sum b_n$ be convergent. Then $\sum (a_n + b_n)$ and $\sum c.a_n$ are convergent.

Proof: This can be checked easily directly from the definition; it is in effect the same proof that the sum of two convergent sequences is convergent etc

The Comparison Test

We have already used the Monotone Convergence Theorem in studying simple series. In fact it is a lot more useful. When we know the behaviour of some simple series, we can deduce many more results by comparison as follows.

Theorem: Assume that $0 \leq a_n \leq b_n$ for all n, and suppose that $\sum b_n$ is convergent. Then a_n is convergent.

Proof: Define

$$S_n = a_1 + a_2 + \cdots + a_n,$$
$$T_n = b_1 + b_2 + \cdots b_n.$$

Then by hypothesis, $0 \leq S_n \leq T_n$. Since $\{T_n\}$ is a convergent sequence, it is a bounded sequence.

In particular, it is bounded above, so there is some K such that $T_n \leq K$ for all n. Thus $S_n \leq K$ for all n, so $\{S_n\}$ is a sequence that is bounded above; since $a_n \geq 0$, $S_{n+1} = S_n + a_n \geq S_n$ and $\{S_n\}$ is an increasing sequence. Thus by the Monotone Convergence Theorem, it is a convergent sequence.

Example: Let $a_n = \dfrac{2n}{3n^3 - 1}$ and let $b_n = \dfrac{1}{n^2}$. Then $\sum a_n$ is convergent

For $n \geq 1$, $n^3 \geq 1$, so $3n^3 - 1 \geq 2n^3$. Thus

$$a_n = \frac{2n}{3n^3 - 1} \leq \frac{2n}{2n^3} = b_n$$

Since we know that $\sum b_n$ is convergent, so is $\sum a_n$.

Remark: The conclusions of *Theorem* remain true even if we only have $a_n \leq b_n$ eventually; for if it holds for $n \geq N$, we replace the inequality by

$$S_n \leq T_n + a_1 + a_2 + \cdots + a_N$$

and this then holds for all n.

Example: Let $a_n = \dfrac{\log n}{n}$, and compare with $b_n = \dfrac{1}{n}$.

Solution: Note that if $n \geq 3$, then $\log n > 1$. We can thus use the "eventually" form of the comparison test; we have

$$a_n = \frac{\log n}{n} > \frac{1}{n} = b_n,$$

We deduce divergence, for if $\sum a_n$ were convergent, it follows that that $\sum b_n$ was convergent.

Limiting form of the Comparison Test

Corollary: Suppose that $a_n > 0$ and $b_n > 0$, and that there is some constant k such that $\lim_{n\to\infty} \frac{a_n}{b_n} = k > 0$. Then $\sum a_n$ is convergent iff $\sum b_n$ is convergent.

Proof: Assume first that $\sum b_n$ is convergent. Since $a_n/b_n \to k$ as $n \to \infty$, eventually (take $\in = k > 0$), we have $a_n \leq 2kb_n$, which is convergent. Hence $\sum a_n$ is convergent. To get the converse, note that $b_n/a_n \to 1/k$ as $n \to \infty$, so we can use the same argument with a_n and b_n interchanged.

Example: Let $a_n = \frac{n}{n^2+1}$ and let $b_n = \frac{1}{n}$. Then $\sum a_n$ is divergent by the limiting form of the comparison test.

Solution: Note that the terms are all positive, so we try to apply the limiting form of the comparison test directly.

$$\frac{a_n}{b_n} = \frac{n}{n^2+1} \cdot \frac{n}{1} = \frac{n^2}{n^2+1} \to 1 \text{ as } n \to \infty$$

Since the limit is non-zero, the use of the limiting form of the comparison test is valid, and we see that $\sum a_n$ is divergent.

This is all very well, but as with the "new sequences from old" programme, we need a few reference sequences before we can get further. One set is the geometric series, which we have already met.

Exercise: Let a_n = and let $b_n = \left(\frac{2}{3}\right)^n$. Use the limiting form of the comparison test to show that $\sum a_n$ is convergent. We also know about $\sum 1/n\alpha$, at least when $\alpha \geq 2$ when it converges, by comparison with $\sum 1/n^2$, and when $\alpha \leq 1$ when it diverges, by comparison with $\sum 1/n$.

Proposition: The sum $\sum \frac{1}{n^{\alpha}}$ is convergent when $\alpha > 1$.

Solution: Assume $\alpha > 1$; we estimate the partial sums. Since $1/n^{\alpha} > 0$, clearly $\{S_n\}$ is an increasing sequence.

Let $S_n = 1 + \frac{1}{2^{\alpha}} + \cdots + \frac{1}{n^{\alpha}}$, and consider the graph of $y = 1/x^{\alpha}$, noting that y is a decreasing function of x (which is where we use that fact that $\alpha > 1$). From a diagram which is essentially the same as that of Fig, shaded area $\int_1^n \frac{dx}{x^{\alpha}}$

$$\frac{1}{2^{\alpha}} + \ldots + \frac{1}{n^{\alpha}} < \left[\frac{1}{1-\alpha}\frac{1}{x^{\alpha-1}}\right]_1^n$$

$$S_n - 1 < \frac{1}{\alpha-1}\left(1 - \frac{1}{n^{\alpha-1}}\right)$$

$$S_n < \frac{1}{\alpha-1} + 1$$

Thus the sequence of partial sums is bounded above, and the series converges.

The Ratio

Theorem: Let $\sum a_n$ be a series, and assume that lim $\frac{|a_n + 1|}{|a_n|} \to r$ as $n \to \infty$. Then if $r < 1$, the series is convergent, if $r > 1$, the series is divergent, while if $r = 1$, the test gives no information.

Example: Let $a_n = \dfrac{2^n (n!)^2}{(2n)!}$. Then $\sum a_n$ is convergent.

Solution: We look the ratio of adjacent terms in the series (of positive terms).

$$\frac{(n+1)^{st}\text{ term}}{n^{th}\text{ term}} = \frac{a_{n+1}}{a_n} = \frac{2^{n+1}(n+1)!(n+1)!}{(2n+2)!}\frac{(2n)!}{2^n n!n!}$$

$$= \frac{2(n+1)^2}{(2n+2)(2n+1)} = \frac{n+1}{2n+1} \to \frac{1}{2} \text{ as } n \to \infty.$$

Since the ratio of adjacent terms in the series tends to a limit which is < 1, the series converges by the ratio test.

Absolute and Conditional Convergence

So far most of our work has been with series all of whose terms are positive. There is a good reason for this; there is very little we can say about series with mixed signs. Indeed there is just one useful result at this level, which is the topic of this section.

The easiest case occurs when the series really can be thought of as a series of positive terms.

Definition: The series $\sum_{n=1}^{\infty} a_n$ is absolutely convergent iff the series $\sum_{n=1}^{\infty} | a_n |$ is convergent.

Definition: The series $\sum_{n=1}^{\infty} a_n$ is conditionally convergent if and only if the series $\sum_{n=1}^{\infty} a_n$ is convergent but not absolutely convergent.

Example: Show that the series $\displaystyle\sum_{n=1}^{\infty} \frac{(-1)^{n+1}}{n^2}$ is absolutely convergent.

Solution: We have $\sum_{n=1}^{\infty}\left|\frac{(-1)^{n+1}}{n^2}\right| = \sum_{n=1}^{\infty}\frac{1}{n^2}$ and this is convergent. So $\sum_{n=1}^{\infty}\frac{(-1)^{n+1}}{n^2}$ is absolutely convergent.

Note:We choose to work with the sign $(-1)^{n+1}$ rather than $(-1)^n$ simply for tidiness; it is usual to start a series with a positive term, so the coefficient of a_1 is chosen to be +. Thus that of a_2 must be - etc if the series is to alternate. Our interest in absolutely convergent series starts by observing that they are in fact all convergent. Indeed this is the easiest way to show a series is convergent if the terms are not all positive.

Proposition: An absolutely convergent series is convergent.

Proof: Assume that $\sum_{n=1}^{\infty} a_n$ is absolutely convergent, and define

$$a_n^+ = \begin{cases} a_n \text{ if } a_n > 0 \\ 0 \text{ if } a_n \le 0 \end{cases} \quad \text{and} \quad a_n^- = \begin{cases} |a_n| & \text{if } a_n < 0 \\ 0 & \text{if } a_n \ge 0 \end{cases}$$

The point of this definition is that

$0 \le a_n + \le |a_n|$ and $0 \le a_n - |a_n|$ for all n,

so we have two new series of positive terms, while

$$|a_n| = a_n^+ + a_n^- \quad \text{and} \quad a_n = a_n^+ - a_n^-.$$

Using equation to compare with the convergent series $\sum_{n=1}^{\infty} |a_n|$, we see that each of

$$\sum_{n=1}^{\infty} a_n^+ \text{ and } \sum_{n=1}^{\infty} a_n^-$$

is a convergent series of positive terms. Thus

$$\sum_{n=1}^{\infty} a_n = \sum_{n=1}^{\infty} a_n + -a_n^- = \sum_{n=1}^{\infty} a_{n^+} + \sum_{n=1}^{\infty} a_n^-$$

is also convergent using.

This gives one way of proving that a series is convergent even if the terms are not all positive, and so we can't use the comparison test directly. There is essentially only one other way, which is a very special, but useful case known as Leibniz theorem, or the theorem on alternating signs, or the alternating series test. We give the proof because the argument is so like the proof of the convergence of the ratio of adjacent terms in the Fibonacci series.

Warning:Note how we usually talk about the "Fibonacci series", even though it is a *sequence* rather than a series. Try not to be confused by this popular but inaccurate usage.

Leibniz Theorem

Theorem: Let $\{a_n\}$ be a decreasing sequence of positive terms such that $a_n 0$ as $n \to \infty$. Then the series

$$\sum_{n=1}^{\infty}(-1)^{n+1} a_n \text{ is convergent.}$$

Proof: Write S_n for the nth partial sum of the series

$\sum_{n=1}^{\infty}(-1)^{n+1} a_n$. We show this sequence has the same type of oscillating behaviour that the corresponding sequence of partial sums in the Fibonacci example. By definition, we have

$$s_{2n+1} = a_1 - a_2 + a_3 - ... + a_{2n-1} - a_{2n} + a_{2n+1}$$

$$s_{2n-1} = a_1 - a_2 + a_3 - ... + a_{2n-1}$$

and so, subtracting, we have

$$s_{2n+1} = s_{2n+1} - a_{2n} + a_{2n+1}.$$

Since $\{a_n\}$ is a decreasing sequence, $a_{2n} > a_{2n+1}$ and so $s_{2n+1} < s_{2n-1}$. Thus we have a decreasing sequence

$$s_1 > s_3 \quad > s_5 > ... > s_{2n-1} > s_{2n+1} > ....$$

Similarly $s_{2n} > s_{2n-2}$ and we have an increasing sequence

$$s_2 < s_4 < s_6 ... < s_{2n-2} < s_{2n} < ....$$

Also

$$s_{2n+1} = s_{2n} + a_{2n+1} > s_{2n}$$

Thus

$$s_2 < s_4 < s_6 ... < s_{2n-2} < s_{2n} < s_{2n+1} < s_{2n-1} < ... s_5 < s_3 < s_1$$

and the sequence s_1, s_3, s_5, ... is a decreasing sequence which is bounded below (by s_2), and is convergent to a(say). Similarly s_2, s_4, s_6, ... is an increasing sequence which is bounded above (by s_1), and s is convergent to b (say) also

$$s_{2n+1} - s_{2n} = a_{2n+1}$$

and so letting $n \to \infty$

$$\alpha - \beta = 0$$

So $\alpha = \beta$, and all the partial sums are tending to , so the series converges.

Example: Show that the series $\sum_{n=1}^{\infty} \frac{(-1)^{n+1}}{n}$ is conditionally convergent.

Solution: We have $\sum_{n=1}^{\infty} \left| \frac{(-1)^{n+1}}{n} \right| = \sum_{n=1}^{\infty} \frac{1}{n}$ and this is divergent; thus the series is *not* absolutely convergent.

We show that this series is still convergent, and so is conditionally convergent.

Write $a_n = 1/n$, so $a_n > 0$, $a_{n+1} < a_n$ and $a_n \to 0$ as $n \to \infty$. Thus all the conditions of Leibniz's theorem are satisfied, and

so the series $\sum_{n=1}^{\infty} \frac{(-1)^{n+1}}{n}$ is convergent.

Re-arranging an Absolutely convergent Series

Proposition: Let $\sum_{n=1}^{\infty} a_n$ be an absolutely convergent series and suppose that $\{b_n\}$ is a re-arrangement of $\{a_n\}$. Then $\sum_{n=1}^{\infty} b_n$ is convergent, and

$$\sum_{n=1}^{\infty} b_n = \sum_{n=1}^{\infty} a_n .$$

Warning: It is not useful to re-arrange conditionally convergent series. There is a result which is an extreme form of this: Pick $x \in \mathbb{R}$, and let a_n be a conditionally convergent series. then there is a re-arrangement $\{b_n\}$ of $\{a_n\}$ such that $\sum_{n=1}^{\infty} b_n = x$! In other words, we can re-arrange to get any answer we want!

An Estimation Problem

This section shows how we can use a lot of the earlier ideas to produce what is often wanted in practice—results which are an approximation, together with an indication of how good the approximation is. Find how accurate the approximation obtained by just using the first ten terms is, to,

$$\sum_{n=1}^{\infty} \frac{1}{n^2} .$$

Writing this out geometrically gives the statement:

$$\sum_{n=N+1}^{\infty} \frac{1}{n^2} \le \int_N^M \frac{dx}{x^2} \le \sum_{n=N}^{M-1} \frac{1}{n^2}$$

We can evaluate the middle integral:

$$\int_N^M \frac{dx}{x^2} = \left[-\frac{1}{x}\right]_N^M = \frac{1}{N} - \frac{1}{M}$$

For convenience, we define

$$S = \sum_{n=1}^{\infty} \frac{1}{n^2} \text{ and } S_N = \sum_{n=1}^{N} \frac{1}{n^2} .$$

We can now express our inequality in these terms:

$$S_M - S_N \le \frac{1}{N} - \frac{1}{M} \le S_{M-1} - S_{N-1}$$

Next, let $M \to \infty$, so $S_M \to S$, and $1/M \to 0$. We have

$$S - S_N \leq \frac{1}{N} \leq S - S_{N-1}$$

Replacing N by $N + 1$, gives another inequality, which also holds, namely

$$S - S_{N+1} \leq \frac{1}{N+1} \leq S - S_N,$$

and combining these two, we have

$$S - S_{N+1} \leq \frac{1}{N+1} \leq S - S_N \leq \frac{1}{N} \leq S - S_{N-1}.$$

In particular, we have both upper and lower bounds for $S - S_N$, as

$$\frac{1}{N+1} \leq S - S_N \leq \frac{1}{N}$$

To make the point that this is a useful statement, we now specialise to the case when $N = 10$. Then

$$\frac{1}{11} \leq S - S_{10} \leq \frac{1}{10}$$

or
$$0 \leq S - \left(S_{10} + \frac{1}{11} \right) \leq \frac{1}{10} - \frac{1}{11} = \frac{1}{110}.$$

Our conclusion is that although S_{10} is not a very good approximation, we can describe the error well enough to get a much better approximation.

Power Series

Power Series and the Radius of Convergence

The idea of writing $f(x) = P_n(x) + R_n(x)$, to express a function in terms of its Taylor polynomial, together with a remainder. For some functions, the remainder $R_n(x) \to 0$ as $n \to \infty$ for each fixed x. We now recognise this as showing that certain series converge.

We have more effective ways of showing that such a series converges we can use that ratio test. But note that such a test will only show that a series converges, not that it converges to the function used to generate it in the first place.

Proposition: The following series converge for all values of x to the functions shown:

$$e^x = 1 + x + \frac{x^2}{2!} + \frac{x^2}{2!} + \ldots \frac{x^n}{n!} + \ldots$$

$$\sin x = x - \frac{x^3}{3!} + \frac{x^5}{5!} + \ldots + (-1)^{n+1} \frac{x^{2n+1}}{(2n+1)!} + \ldots$$

$$\cos x = 1 - \frac{x^2}{2!} + \frac{x^4}{4!} + \ldots + (-1)^n \frac{x^{2n}}{2n!} + \ldots$$

$$\sinh x = x + \frac{x^3}{3!} + \frac{x^5}{5!} + \ldots + \frac{x^{2n+1}}{(2n+1)!} + \ldots$$

$$\cosh = 1 + \frac{x^2}{2!} + \frac{x^4}{4!} + \ldots + \frac{x^{2n}}{2n!} + \ldots$$

These are all examples of the subject of this section; they are real power series, which we can use to define functions. The corresponding functions are the best behaved of all the classes of functions; indeed are as well behaved as could possibly be expected. We shall see in this section that this class of functions are really just "grown up polynomials", and that almost any manipulation valid for polynomials remains valid for this larger class of function.

Definition: A real power series is a series of the form $\sum a_n x^n$, where the a_n are real numbers, and x is a real variable. We are thus dealing with a whole collection of series, one for each different value of x. Our hope is that there is some coherence; that the behaviour of series for different values of x are related in some sensible way.

Example: The geometric series $\sum_{n=0}^{\infty} x^n$ is another example of a power series we have already met. We saw this series is convergent for all x with $|x| < 1$. It turns out that a power series is usually best investigated using the ratio test, Theorem. And the behaviour of power series is in fact very coherent.

Radius of Convergence

Theorem: Suppose $\sum a_n x^n$ is a power series. Then one of the following happens:

1. $\sum a_n x^n$ converges only when $x = 0$; or
2. $\sum a_n x^n$ converges absolutely for all x; or
3. There is some number $R > 0$ such that $\sum a_n x^n$ converges absolutely for all x with $|x| < R$, and diverges for all x with $|x| > R$.

No statement is made in the third case about what happens when $x = R$.

Definition: The number R described above is called the radius of convergence of the power series. By allowing $R = 0$ and $R = \infty$, we can consider *every* power series to have a radius of convergence.

Thus every power series has a radius of convergence. We sometimes call the interval $(-R, R)$, where the power series is guaranteed to converge, the interval of convergence. It is characterised by the fact that the series converges (absolutely) inside this interval and diverges outside the interval.

1. The word "radius" is used, because in fact the same result is true for complex series, and then we have a genuine circle of convergence, with convergence for all (complex) z with $|z| < R$, and guaranteed divergence whenever $|z| > R$.
2. Note the power of the result; we are guaranteed that when $|x| > R$, the series diverges; it can't even converge "accidentally" for a few x's with $|x| > R$. Only on the circle of convergence is there ambiguity.

This regularity of behaviour makes it easy to investigate the radius of convergence of a power series using the ratio test.

Example: Find the radius of convergence of the series $\sum \frac{nx^n}{2^{n+1}}$.

Solution: Recall that the ratio test only applies to series of positive terms, so we look at the ratio of the moduli.

$$\frac{\left|(n+1)^{st} \text{ term}\right|}{\left|n^{th} \text{ term}\right|} = \frac{(n+1)|x|^{n+1}}{2^{n+2}} \frac{2^{n+1}}{n|x|^n}$$

$$= \frac{(n+1)|x|}{n.2} \to \frac{|x|}{2} \text{ as } n \to \infty .$$

Thus the given series diverges if $|x| > 2$ and converges absolutely if $|x| < 2$. Hence it has radius of convergence 2.

Example: Find the radius of convergence of the series $\sum \frac{(-1)^n n!x^n}{n^n}$.

Solution: This one is a little more subtle than it looks, although we have met the limit before. Again we look at the ratio of the moduli of adjacent terms.

$$\frac{\left|(n+1)^{st} \text{ term}\right|}{\left|n^{th} \text{ term}\right|} = \frac{(n+1)!|x|^{n+1}}{(n+1)^{n+1}} \frac{n^n}{n!|x|^n} = (n+1)\cdot|x|\cdot\frac{n^n}{(n+1)^{(n+1)}}$$

$$= \frac{n^n}{(n+1)^n}|x| = \left(\frac{n}{n+1}\right)^n |x|$$

$$= \frac{|x|}{\left(\frac{n+1}{n}\right)^n} \frac{|x|}{e}$$

as $n \to \infty$

Here we have of course used the result about e given in Section to note that

$$\left(1+\frac{1}{n}\right)^n \to e \text{ as } n \to \infty .$$

Thus the given series diverges if $|x| > e$ and converges absolutely (and so of course converges) if $|x| < e$. Hence it has radius of convergence e.

Exercise: Find the radius of convergence of the series

$$\sum_{n=0}^{\infty} \frac{x^n}{n^2+1}$$

We noted that the theorem gives no information about what happens when $x = R$, i.e. on the circle of convergence. There is a good reason for this — it is quite hard to predict what happens. Consider the following power series, all of which have radius of convergence 2.

$$\sum_{n=0}^{\infty} \frac{x^n}{2^n} \quad \sum_{n=1}^{\infty} \frac{x^n}{n2^n} \quad \sum_{n=1}^{\infty} \frac{x^n}{n^2 2^n} .$$

The first is divergent when $x = 2$ and when $x = -2$, the second converges when $x = -2$, and diverges when $x = 2$, while the third converges both when $x = 2$ and when $x = -2$. These results are all easy to check by direct substitution, and using Theorem.

Representing Functions by Power Series

Once we know that a power series has a radius of convergence, we can use it to define a function.

Suppose the power series $\sum a_n x^n$ has radius of convergence $R > 0$, and let $I = (-R, R)$. We now define a function f on this open interval I as follows:

$$f(x) = \sum_{n=1}^{\infty} a_n x^n \quad \text{for } x \in I.$$

In fact all of what we say below remains true when $R = \infty$, provided we interpret the open interval I as $\mathbb{R}$.

Theorem: Let $\sum a_n x^n$ be a power series with radius of convergence $R > 0$.

Let I be the open interval $(-R, R)$, and define $f(x) = \sum a_n x^n$ for $x \in I$. Then $\sum n a_n x^{n-1}$ has radius of convergence R, f is differentiable on I, and

$$f'(x) = \sum_{n=1}^{\infty} n a_n x^{n-1} \quad \text{for } x \in I.$$

We summarise this result by saying that we can differentiate a power series term – by - term everywhere inside the circle of convergence. If $R = \infty$, then this can be done for all x.

Proof: Quite a lot harder than it looks; we need to be able to re-arrange power series, and then use the Mean Value Theorem to estimate differences, and show that even when we add an infinite number of errors, they don't add up to too much.

Corollary: Let f and I be defined. Then f has an indefinite integral defined on I, given by

$$G(x) = \sum_{n=0}^{\infty} \frac{a_n}{(n+1)} x^{n+1} \quad \text{for } x \in I.$$

Other Power Series

We now derive some further power series to add to the collection described. Starting with the geometric series

$$\frac{1}{1-x} = 1 + x + x^2 + x^3 + \ldots + x^n + \ldots \quad \text{for } |x| < 1,$$

we replace x by $-x$ to get

$$\frac{1}{1+x} = 1 - x + x^2 - x^3 + \ldots + (-1)^n x^n + \ldots \quad \text{for } |x| < 1.$$

Integrating both sides then gives

$$\log(1 + x) = K + x - \frac{x^2}{2} + \frac{x^3}{3} + \ldots + (-1)^n \frac{x^{n+1}}{n+1} + \ldots \quad \text{for } |x| < 1,$$

where K is a constant of integration. Evaluating both sides when $x = 0$ shows that $K = 0$, and so we get the series

$$\log(1 + x) = \sum_{n=0}^{\infty} \frac{(-1)^n x^{n+1}}{n+1} \quad \text{valid for } |x| < 1$$

We return to equation, and replace x by $-x^2$ to get

$$\frac{1}{1+x^2} = 1 - x^2 + x^4 - x^6 + \ldots + (-1)^n x^{2n} + \ldots \quad \text{for } |x| < 1.$$

Again integrating both sides, we have arctan $(x) = K + x -$ $+\ldots+ (-1)^n \frac{x^{2n+1}}{2n+1}$ $+\ldots$ for $|x| < 1$, where K is a constant of integration. Again putting $x = 0$ shows that $K = 0$, and so

$$\arctan(x) = \sum_{n=0}^{\infty} \frac{(-1)^n x^{2n+1}}{2n+1} \quad \text{valid for } |x| < 1.$$

Example: Find the radius of convergence R of the power series

$$\frac{x^2}{2} - \frac{x^3}{3.2} + \frac{x^4}{4.3} - \frac{x^5}{5.4} .. + \frac{(-1)^n x^n}{n(n-1)} + \ldots$$

By differentiation or otherwise, find the sum of the series for $|x| < R$. [You may assume, without proof, that

$$\int \log(1+x)\,dx = K + x\log(1 + x) - x + \log(1 + x),$$

for some constant of integration K.]

Solution: Apply the ratio test to the given power series. Then

$$\left|\frac{a_n+1}{a_n}\right| = \left|\frac{x^{n+1} n(n-1)}{x^n n(n+1)}\right| \to |x| \quad \text{as } n \to \infty .$$

Thus the new series has radius of convergence 1. Denote its sum by $f(x)$, defined for $|x| < 1$.

Inside the circle of convergence, it is permissible to differentiate term - by - term, and thus $f'(x) = \log(1 + x)$ for $|x| < 1$, since they have the same power series. Hence

$$f(x) = \int \log(1+x)\,dx = K + x\log(1+x) - \int \frac{x+1-1}{1+x}\,dx$$

$$= K + x\log(1+x) - x + \log(1+x).$$

Putting $x = 0$ shows that $K = 0$ and so $f(x) = (1 + x)\log(1 + x) - x$.

We have now been able to derive a power series representation for a function without working directly from the Taylor series, and doing the differentiations — which can often prove very awkward. Nevertheless, we have still found the Taylor coefficients.

Proposition: Let $\sum a_n x^n$ be a power series, with radius of convergence $R > 0$, and define

$$f(x) = \sum_{n=0}^{\infty} a_n x^n \quad \text{for } |x| < R.$$

Then $a_n = \dfrac{f^{(n)}(0)}{n!}$, so the given series *is* the Taylor (or Maclaurin) series for f

Proof: We can differentiate n times and we still get a series with the same radius of convergence. Also, calculating exactly as in the start of Section, we see that the derivatives satisfy $f^{(n)}(0) = n!a_n$, giving the uniqueness result.

Example: Let $f(x) = \dfrac{1}{1-x^3}$. Calculate $f^{(n)}(0)$.

Solution: We use the binomial theorem to get a power series expansion about 0,

$$\frac{1}{1-x^3} = 1 + x^3 + x^6 + x^9 + \ldots + x^{3n} + \ldots \quad \text{valid for } |x| < 1.$$

We now read off the various derivatives. Clearly $f^{(n)}(0) = 0$ unless n is a multiple of 3, while $f^{(3k)}(0) = (3k)!$

Power Series or F unction

We have now seen that when a power series is used to define a function, then that function is *very* well behaved, and we can manipulate it by manipulating, in the obvious way, the corresponding power series. However there are snags. A function has *one* definition which works everywhere it makes sense (at least for simple functions), whereas the power series corresponding to a function depends also on the point about which the expansion is happening. An example will probably make this clearer than further discussion.

Example: Give the power series expansions for the function $f(x) = \dfrac{1}{1-x}$.

Solution: We can already do this about 0 by the Binomial Theorem; we have:

$$\frac{1}{1-x} = 1 + x + x^2 + x^3 + \ldots + x^n + \ldots \quad \text{for } |x| < 1.$$

To expand about a different point, e.g. about 3, write $y = x - 3$. Then

$$\frac{1}{1-x} = \frac{1}{1-(y+3)} = = \frac{1}{-2-y} = -\frac{1}{2}\cdot\frac{1}{1-y/2}\text{,}$$

and again using the Binomial Theorem on the last representation, we have

$$\frac{1}{1-x} = -1\,\frac{1}{2}\,(1 + y/2 + y^2/4 + y^3/9 + \ldots + y^n/2^n + \ldots) \text{ for } |y/2| < 1.$$

Writing this in terms of x gives

$$\frac{1}{1-x} = -\frac{1}{2}\left(1 + \frac{x-3}{2} + \frac{(x-3)^2}{4} + \ldots + \frac{(x-3)^n}{2^n} + \ldots\right)$$

for $|x - 3| < 2$.

It should be no surprise that this is the Taylor series for the same function about the point 3. And it is in fact not an accident that the radius of convergence of the new series is 2. More investigation (quite a lot more - mainly for complex functions) shows the radius of convergence is always that of the *largest* circle that can be fitted into the domain of definition of the function. And that is why it is of interest to sometimes consider power series as complex power series. The power series expansion for $(1 + x^2)^{-1}$ has radius of convergence 1. This seems implausible viewed with real spectacles, but totally explicable when it is realised that the two points i and $-i$ are stopping the expansion from being valid in a larger circle.

The Definition and Limit Method

The derivative of a function expresses its rate of change with respect to an independent variable. The derivative is also the slope of the tangent line to the curve. As you might recall. Many students start out differentiating with the Limit Method. The Limit Method utilizes the definition of a derivative of a function. The Limit Method, as seen below, is imperative because some problems may not work or may take longer with the differentiation rules.

$$\lim_{c \to 0} \frac{f(x+c) - f(x)}{c}$$

(Where c denotes a small distance on the x-axis usually written as the lower case delta sign next to an x).

Chapter 8

Derivative Equations

Rate of Change

The derivative of a function at a chosen input value describes the best linear approximation of the function near that input value. For a real-valued function of a single real variable, the derivative at a point equals the slope of the tangent line to the graph of the function at that point. In higher dimensions, the derivative of a function at a point is a linear transformation called the linearization. The process of finding a derivative is called differentiation.

The fundamental theorem of calculus states that differentiation is the reverse process to integration. The mathematical relationship between an independent and dependent dimension. The relationship expressed in the form of a mathematical function defines a situation for a given set of conditions and properties. Since we allowed the independent dimension to change, it was no longer a fixed quantity but became a variable.

The graph of the function reflected a horizontal change of the independent variable with a vertical change in the dependent dimension. This graph allows us to visualize how a situation changes with respect to a change in the conditions.

To find out how much the function, $f(x)$, changes between two points x_1 and x_2, we simply enter in the two values for the independent variable x and then calculate the difference between the dependent variable, *f*, for those given conditions. Remember that a variable is nothing more than a dimension that is allowed to change or take on any value. Thus, from x_1 to x_2, the change in the independent variable, referred to as Δx is:

$$\Delta x = x_1 - x_2$$

The corresponding change in the dependent variable , referred to as Δf is:

$\Delta f = f(x_2) - f(x_1)$ Δx refers to an interval over which we are analyzing the change in the dependent dimension, *f*.. The second point x_2 can be written in terms of the first point x_1 plus the change in the variable, Δx

$$x_2 - x_1 = \Delta x$$
$$\therefore x_2 = x_1 + \Delta x$$

Therefore the change in the dependent variable over an interval $x_1 x_2$ from $\Delta f = f(x_2) - f(x_1)$ which can also be written as:

$$\Delta f = f(x_1 + \Delta x) - f(x_1)$$

where $\Delta x = x_2 - x_1$

For example the change in the function $f(x) = 3 \cdot x^3$ $x = 3$ to $x = 5$, where

Δx is 5 – 3 = 2 is:

$$\Delta f = f(x_1 + \Delta x) - f(x_1)$$
$$\Delta f = f(3+2) - f(3)$$
$$\Delta f = 3.5^3 - 3.3^3 = 375 - 81$$
$$\Delta f = 294$$

Average Rate of Change

We have learned that a change in the independent variable is defined as $\Delta x = x_2 - x_1$, and the corresponding change in the dependent variable over this interval is $\Delta f = f(x_1 + \Delta x) - f(x_1)$ The question we now must ask ourselves is how can we measure the relative change of the dependent variable with respect to the independent variable? In other words how can we calculate how much more or less Δf changed compared to Δx .

To calculate how much more $f(x)$ changed over an interval from x_1, x_2, we simply divide the change in f over the change in x for the interval. Thus we divide, Δf by the interval over we which are evaluating it, Δx which is equal to $x_2 - x_1$. Thus the relative change of f with respect to x over an interval Δx is defined as:

$$\frac{\Delta f}{\Delta x} = \frac{f(x_2) - f(x_1)}{x_2 - x_1}$$

While this expression may seem rather simple, it does require some explanation. By dividing the change in f by the change in x what we are doing is calculating how much more f changed for a given change in x. For example in the function, $f(x) = 3.x^3$, when x changed from 3 to 5, f changed from 81 to 375. Over this interval of from $x = 3$ to $x = 5$, the Δf was 294.

Thus the relative change in f with respect to a change in the independent variable x is:

$$\frac{\Delta f}{\Delta x} = \frac{f(x_2) - f(x_1)}{x_2 - x_1}$$

The value of 147 tells us that f changes 147 times faster than x over that interval of Δx from $x = 3$ to $x = 5$ only. Thus for each unit change in x, $\Delta x = 1$, the corresponding change in f is 147. We can therefore define the rate of change of a function with respect to its independent variable to be:

$$\frac{\Delta f}{\Delta x} = \frac{f(x_1 + \Delta x) - f(x_1)}{\Delta x}$$

The value, called the rate of change of the function, refers to how much more or less $f(x)$ changes for a unit change in x. Itis only valid is for the interval under consideration, $\Delta x = x_2 - x_1$.

As Δf increases, the steepness of the line connecting the two endpoints will increase. Thus, the greater the rate of change of the function, the greater its slope or steepness over the interval under consideration.

Since slope and rate of change are synonymous, then how is rate of change defined for functions whose graphs do not have constant slopes? For example, from x = 9 to x = 12 of $f(x) = 3.x^3$, the change in f is 2997. Thus the rate of change of the function over the interval is:

$$\frac{\Delta f}{\Delta x} = \frac{2997}{3} = 999$$

This value is significantly higher than the rate of change calculated for the previous interval from x = 3 to x = 5.

Since the rate of change of a function can change, then we have to come up with a more refined definition of rate of change. We can define the average rate of change of a function over an interval $\Delta x - x_2 - x_1$, to be equal to

$$\text{Average rate of change} = \frac{\Delta f}{\Delta x} = \frac{f(x_1 + \Delta x) - f(x_1)}{\Delta x}$$

In the next section we will take a closer look at how we can define the exact or instantaneous rate of change of a function.

Instantaneous Rate of Change

In this section we will take a much closer look at rate of change and see how we can define the instantaneous rate of change of a function at any point or value of the independent variable.

Let us begin with a study of the simple $f(x) = 3x$. Functions of the form $f(x) = cx$, where c is a constant, express direct relationships. This is because the value of the function, *f*, is a constant multiple or fraction of the independent variable. Thus f is said to be directly proportional to x. The graph of the function $f(x) = 3x$ looks like. From $x = 1$ to $x = 3$, $\Delta x = 2$, and $\Delta f = 9 -$ $3 = 6$. Thus, over this interval of $\Delta x = 2$, Δf equals 6. The average rate at which f changed with respect to x is by definition,

$\frac{\Delta f}{\Delta x} = \frac{6}{2} = 3$. For each unit change in x, the change in f is 3. This tells us that f is changing three times faster that x is changing over the interval from $x = 1$ to $x = 3$. We can now look at the interval from $x = 2$ to $x = 4$ where Δx equals 2.

$$\Delta f = f(x_1 + \Delta x) - f(x_1)$$
$$\Delta f = f(2+2) - f(2)$$
$$\Delta f = 3.(4) - 3.(2)$$
$$\Delta f = 6$$

Thus

$$\frac{\Delta f}{\Delta x} = \frac{f(x_1 + \Delta x) - f(x_1)}{\Delta x} = \frac{6}{2} = 3$$

Over this interval, the rate of change is the same constant, 3. This leads us to conclude that the rate of change of the function over any interval is a constant, 3. This can be proven by the definition of rate of change:

$$f(x) = 3x$$
$$f(x + \Delta x) = 3(x + \Delta x)$$
$$\frac{\Delta f}{\Delta x} = \frac{f(x+\Delta x) - f(x)}{\Delta x} = \frac{3\cancel{x} + 3\Delta x - 3\cancel{x}}{\Delta x}$$
$$\frac{\Delta f}{\Delta x} = \frac{3\Delta\cancel{x}}{\Delta\cancel{x}} = 3$$

On the other hand, the graph of the function $f(x) = x^2$ is not a straight line. Unlike a line, the rate of change of $f(x)$ is not the same constant over any interval Dx. To define the rate of change for the function we will have to derive a more precise way of defining rate of change of function. To do this we will analyze the function over small intervals of Dx

The first point to notice is that the rate of change of the function varies with x. When x is small, f does not change that much as compared to how much it changes when x is large. We can conclude that the rate of change of f with respect to x, is not constant over any interval, Dx, but varies with x. To begin our analysis let us divide up the graph into intervals of D$x = 2$

and study what is happening in each of these intervals separately. From $x = 0$ to $x = 2$, $Dx = 2$, the change in f, D*f, is* $f(2) - f(0) = 4$. From $x = 6$ to $x = 8$, D$x = 2$, the change in *f*, D*f, is* $f(8) - f(6) = 28$.

Clearly, as x increases the rate of change of the function is increasing and is not a constant as in our study of the line where the function changed at the same rate as x increased.

Returning back to the graph of $f(x) = x^2$; we calculated that from $x = 0$ *to* $x = 2$ the change in f was 4. We can then conclude that from $x = 0$ to $x = 2$ the average rate of change of the function over that particular interval is 2 or:

$$\frac{\Delta f}{\Delta x} = \frac{4}{2} = 2$$

Remember it is called average because this rate of change is only valid from $x = 0$ to $x = 2$. Now let us consider the interval from $x = 2$ to $x = 4$ where once again the change in x, Dx, is 2. The change in f of the graph is equal to:

$$\begin{aligned}\Delta f &= f(x + \Delta x) - f(x) \\ &= f(2+2) - f(2) \\ &= f(4) - f(2) \\ &= 16 - 4 \\ &= 12\end{aligned}$$

Thus the average rate of change of the function over this interval is equal to

$$\frac{\Delta f}{\Delta x} = \frac{12}{2} = 6$$

This value is greater than the value we observed from $x =$ 0 to $x = 2$. This implies that from $x = 2$ to $x = 4$, f(x) is increasing at a greater rate than from $x = 0$ to $x = 2$. This is despite the fact that in both cases the D$x = 2$.

The rate of change is therefore not constant over the but is increasing with x.

When we say the rate of change of $f(x)$ from $x = 2$ to $x = 4$ is 6, it is only the average value for the given interval since it assumes rate of change is constant over that interval.

Let us move on to the next interval of $x = 4$ to $x = 6$. Once again the change in x or Dx equals 2 but the corresponding change in $f(x)$ is not the same as before.

$$\begin{aligned}\Delta f &= f(x+\Delta x) - f(x)\\ &= (x+\Delta x)^2 - (x)^2\\ &= x^2 + 2x\Delta x + \Delta x^2 - x^2\\ &= 2x\Delta x + \Delta x^2\end{aligned}$$

For x equal to 4 and Dx = *2*, the change in the function over this interval is

$$\begin{aligned}\Delta f &= 2x\Delta x + \Delta x^2\\ \Delta f &= 2(4)(2) + (2)^2\\ \Delta f &= 20\end{aligned}$$

Note that this corresponds to the same value we would get by:

$$\begin{aligned}\Delta f &= f(6) - f(4)\\ &= 6^2 - 4^2\\ &= 36 - 16\\ &= 20\end{aligned}$$

The average rate of change over this interval is therefore:

$$\frac{\Delta f}{\Delta x} = \frac{20}{2} = 10$$

This is still larger than the rate of change for the previous interval which was 6. Remember rate of change , by definition, refers to how much the function changes with respect to a change in the independent variable. The steepness or slope of the line over the interval provides a geometric understanding for this concept.

The idea being we need to analyze our graph over a small interval, Dx, to see what exactly is going on at each instant the function is changing. Here is where we begin our study of Calculus. We break down and freeze a changing situation into an infinite series of actions and analyzing what is going on in each individual actions.

Basic Properties of Derivative

We continue our study of classes of functions which are suitably restricted. Again we are passing from the general to the particular. The next most particular class of function we study after the class of continuous functions is the class of differentiable functions. We discuss the definition, show how to get "new functions from old" in what by now is a fairly routine way, and prove that this is a smaller class: that every differentiable function is continuous, but that there are continuous functions that are not differentiable. Informally, the difference is that the graph of a differentiable function may not have any sharp corners in it.

As with continuous functions, our aim is to show that there are many attractive properties which hold for differentiable functions that don't hold in general for continuous functions. One we discuss in some detail is the ease with which certain limits can be evaluated, by introducing l'Hopital's rule. Although we don't prove this, the corresponding results are false for continuous functions. We take the view that much of this material has already been discussed last year, so we move fairly quickly over the basics.

Definition: Let U be an open subset of $\mathbb{R}$, and let f : $U \to \mathbb{R}$. We say that f is differentiable at $a \in U$ iff

$$\lim_{x \to \alpha} \frac{f(x) - f(a)}{x - a}$$

or equivalently,

$$\lim_{h \to 0} \frac{f(a+h) - f(a)}{h}$$

exists.

The limit, if it exists, is written as $f'(a)$.

We say that f is differentiable in U if and only if it is differentiable at each a in U.

Note that the Newton quotient is not defined when $x = a$, nor need it be for the definition to make sense. But the Newton quotient, if it exists, can be extended to be a continuous function at a by defining its value at that point to be $f'(a)$. Note also the emphasis on the existence of the limit. Differentiation is as much about showing the *existence* of the derivative, as *calculating* the value of the derivative.

Example: Let $f(x) = x^3$. Show, directly from the definition, that $f'(a) = 3a^2$.

Solution: This is just another way of asking about particular limits, we must compute

$$\lim_{x \to \alpha} \frac{x^3 - a^3}{x - a} = \lim_{x \to \alpha} \frac{(x-a)(x^2 + xa + a^2)}{x - a} =$$

$$\lim_{x \to \alpha} (x^2 + 2xa + a^2) = 3a^2.$$

Exercise: Let $f(x) = \sqrt{x}$. Show, directly from the definition, that $f'(a) = 1/2\ \sqrt{a}$ when $a \neq 0$. What function do you have to consider in the particular case when $a = 4$?

Just as with continuity, it is impractical to use this definition every time to compute derivatives; we need results showing how to differentiate the usual class of functions, and we assume these are known from last year. Thus we assume the rules for differentiation of sums products and compositions of functions, together with the known derivatives of elementary functions such as sin, cos and tan; their reciprocals sec, *cosec* and cot; and exp and log.

Proposition: Let f and g be differentiable at a, and let k be a constant. Then $k.f$, $f + g$ and fg are differentiable at f. Also, if $g(a) \neq 0$, then f/g is differentiable at a. Let f be differentiable at a, and let g be differentiable at $f(a)$. Then gof is differentiable at a. In addition, the usual rules for calculating these derivatives apply.

Example: Let $f(x) = \tan\left(\dfrac{x^2 - a^2}{x^2 + a^2}\right)$ for $a \neq 0$. Show that f

is differentiable at every point of its domain, and calculate the derivative at each such point.

Solution: This is the same example we considered in. There we showed the domain was the whole of $\mathbb{R}$, and that the function was continuous everywhere. Let $g(x) = \dfrac{x^2 - a^2}{x^2 + a^2}$. Then g is properly defined for all values of x, and the quotient is differentiable, since each term is, and since $x^2 + a^2 \neq 0$ for any x since $a \neq 0$. Thus f is differentiable using the chain rule since $f =$ tanog, and we are assuming known that the elementary functions like tan are differentiable.

Finally to actually calculate the derivative, we have:

$$f'(x) = \sec^2\left(\frac{x^2 - a^2}{x^2 + a^2}\right)\frac{\left(x^2 + a^2\right)\cdot 2x - \left(\left(x^2 - a^2\right)\cdot 2x\right)}{\left(x^2 + a^2\right)^2}.$$

$$= \frac{4a^2 x}{\left(x^2 + a^2\right)^2}\cdot \sec^2\left(\frac{x^2 - a^2}{x^2 + a^2}\right)$$

Exercise: Let $f(x) = \exp\left(\dfrac{1 + x^2}{1 - x^2}\right)$. Show that f is differentiable at every point of its domain, and calculate the derivative at each such point. The first point in our study of differentiable functions is that it is *more* restrictive for a function to be differentiable, than simply to be continuous.

Proposition: Let f be differentiable at a. Then f is continuous at a.

Proof: To establish continuity, we must prove that $\lim\limits_{x \to a} f(x) = f(a)$. Since the Newton quotient is known to converge, we have for $x \neq a$,

$$f(x) - f(a) = \frac{f(x) - f(a)}{x - a}.(x - a) \to f'(a).0 \quad \text{as } x \to a.$$

Hence f is continuous at a.

Example: Let $f(x) = |x|$; then f is continuous everywhere, but *not* differentiable at 0.

Solution: We already know from Example that $|x|$ is continuous. We compute the Newton quotient directly; recall that $|x| = x$ if $x \geq 0$, while $|x| = -x$ if $x < 0$. Thus

$$\lim_{x \to 0+} \frac{f(x) - f(0)}{x - 0} = \lim_{x \to 0+} \frac{x - 0}{x - 0}, \text{ while}$$

$$\lim_{x \to 0-} \frac{f(x) - f(0)}{x - 0} = \lim_{x \to 0-} \frac{-x - 0}{x - 0} = -1.$$

Thus both of the one-sided limits exist, but are unequal, so the limit of the Newton quotient does not exist.

Simple Limits

Our calculus of differentiable functions can be used to compute limits which otherwise prove troublesome.

Proposition: Let f and g be functions such that $f(a) = g(a) = 0$ while $f'(a)$ and $g'(a)$ both exits and $g'(a) \neq 0$. Then

$$\lim_{x \to a} \frac{f(x)}{g(x)} = \frac{f'(a)}{g'(a)}.$$

Proof: Since $f(a) = g(a) = 0$, provided $x \neq a$, we have

$$\frac{f(x)}{g(x)} = \frac{f(x) - f(a)}{g(x) - g(a)} = \frac{f(x) - f(a)}{x - a} \frac{x - a}{g(x) - g(a)}$$

$$\to \frac{f'(a)}{g'(a)} \text{ as } x \to a$$

where the last limit exists, since $g'(a) \neq 0$.

Remark: If $f'(a)$ and $g'(a)$ exist, $\lim_{x \to \alpha} \frac{f(x)}{g(x)}$ computing is easy, since f and g must be continuous at a by Proposition, unless we get an indeterminate form 0/0 or ∞ / ∞ for the formal quotient. In fact l'Hopitals rule helps in both cases, although we need to develop stronger forms.

Example: Show that $\lim\limits_{x\to 0}\frac{3x-\sin x}{x} = 2.$

Solution: Note first that we cannot get the result trivially from, since since $g(a) = 0$ and so we get the indeterminate form 0/0. However, we are in a position to apply the simple form of l'Hopital, since $x' = 1 \neq 0$. Applying the rule gives

$$\lim_{x\to 0}\frac{3x-\sin x}{x} = \lim_{x\to 0}\frac{3-\cos x}{1} = 2.$$

Example: Show that $\lim\limits_{x\to 0}\frac{\sqrt{1+x}-1}{x} = 1/2.$

Solution: Note again that we cannot get the result trivially from 4.16, since this gives the indeterminate 0/0 form, because $g(a) = 0$. However, we are in a position to apply the simple form of l'Hopital, since $x' = 1 \neq 0$. Applying the rule gives

$$\lim_{x\to 0}\frac{\sqrt{1+x}-1}{x} = \lim_{x\to 0}\frac{2^{-1}(1+x)^{-1/2}}{1} = 1/2.$$

Example: Show that $\lim\limits_{x\to 0}\frac{\sin x}{x} = 1.$

Solution: This is spurious because we need the limit to calculate the derivative in the first place, but applying l'Hopital certainly gives the result.

Rolle and the Mean Value Theorem

We can combine our definition of derivative with the Intermediate Value Theorem to give a useful result which is in fact the basis of most elementary applications of the differential calculus.

Like the results on continuous functions, it is a global result, and so needs continuity and differentiability on a whole interval.

Theorem (Rolle's Theorem): Let f be continuous on $[a, b]$, and differentiable on (a, b), and suppose that $f(a) = f(b)$. Then there is some c with $a < c < b$ such that $f'(c) = 0$.

Proof: Since f is continuous on the compact interval $[a, b]$, it has both a global maximum and a global minimum. Assume first that the global maximum occurs at an interior point $c \in (a, b)$. In what follows, we pick h small enough so that $c + h$ always lies in (a, b). Then

If $h > 0$, $\dfrac{f(c+h)-f(c)}{h} \leq 0$, and so

$$\lim_{h\to 0+} \frac{f(c+h)-f(c)}{h} \leq 0,$$

since we know the limit exists.

Similarly, if $h < 0$, $\dfrac{f(c+h)-f(c)}{h} \geq 0$,

and so

$$\lim_{h\to 0+} \frac{f(c+h)-f(c)}{h} \geq 0,$$

since we know the limit exists. Combining these, we see that $f'(c) = 0$, and we have the result in this case. A similar argument applies if, instead, the global minimum occurs at the interior point c. The remaining situation occurs if both the global maximum and global minimum occur at end points; since $f(a) = f(b)$, it follows that f is constant, and any $c \in (a, b)$ will do.

Example: Investigate the number of roots of each of the polynomials

$$p(x) = x^3 + 3x + 1 \quad \text{and} \quad q(x) = x^3 - 3x + 1.$$

Solution: Since $p'(x) = 3(x^2 + 1) > 0$ for all $x \in \mathbb{R}$, we see that p has at most one root; for if it had two (or more) roots there would be a root of $p'(x) = 0$ between them by Rolle. Since $p(0) = 1$, while $p(-1) = -3$, there is at least one root by the Intermediate Value Theorem. Hence p has exactly one root. We have $q'(x) = 3(x^2 - 1)$

= 0 when $x = \pm 1$. Since $q(-1) = 3$ and $q(1) = -1$, there is a root of q between -1 and 1 by the Intermediate Value Theorem. Looking as $x \to \infty$ and as $x \to -\infty$ shows here are three roots of q.

The Mean Value Theorem

Theorem: Let f be continuous on $[a, b]$, and differentiable on (a, b). Then there is some c with $a < c < b$ such that

$$\frac{f(b)-f(a)}{b-a} = f'(c)$$

or equivalently $f(b) = f(a) + (b - a)f'(c)$.

Proof: We apply Rolle to a suitable function; let

$$h(x) = f(b) - f(x) - \frac{f(b)-f(a)}{b-a}(b - x).$$

Then h is continuous on the interval $[a, b]$, since f is, and in the same way, it is differentiable on the open interval (a, b).

Also, $f(b) = 0$ and $f(a) = 0$. We can thus apply Rolle's theorem to h to deduce there is some point c with $a < c < b$ such that $h'(c) = 0$. Thus we have

$$0 = h'(c) = -f'(c) + \frac{f(b)-f(a)}{b-a},$$

which is the required result.

Example: The function f satisfies $f'(x) = \frac{1}{5-x^2}$ and $f(0) =$ 2. Use the Mean Value theorem to estimate f.

Solution: We first estimate the given derivative for values of x satisfying $0 < x < 1$. Since for such x, we have $0 < x^2 < 1$, and so $4 < 5 - x^2 < 5$. Inverting we see that

$$\frac{1}{5} < f'(x) < \frac{1}{4} \text{ when } 0 < x < 1.$$

Now apply the Mean Value theorem to f on the interval [0, 1] to obtain some c with $0 < c < 1$ such that $f(1) - f(0) = f'(c)$. From the given value of $f(0)$, we see that $2.2 < f(1) < 2.25$

Exercise: The function f satisfies $f¢(x)\frac{1}{5+\sin x} =$ and $f(0) = 0$. Use the Mean Value theorem to estimate f (p/2).

Note the "common sense" description of what we have done. If the derivative doesn't change much, the function will behave linearly. Note also that this gives meaning to the approximation

$$f(a + h) \approx f(a) + hf'(a).$$

We now see that the accurate version of this replaces $f'(a)$ by $f'(c)$ for some c between a and $a + h$.

The Cauchy Mean Value Theorem

Theorem: Let f and g be both continuous on $[a, b]$ and differentiable on (a, b). Then there is some point c with $a < c < b$ such that

$$g'(c)(f(b) - f(a)) = f'(c)(g(b) - g(a)).$$

In particular, whenever the quotients make sense, we have

$$\frac{f(b)-f(a)}{g(b)-g(a)} = \frac{f'(c)}{g'(c)}$$

Proof. Let $h(x) = f(x)(g(b) - g(a)) - g(x)(f(b) - f(a))$, and apply Rolle's theorem exactly as we did for the Mean Value Theorem. Note first that since both f and g are continuous on $[a, b]$, and differentiable on (a, b), it follows that h has these properties. Also $h(a) = f(a)g(b) - g(a)f(b) = h(b)$. Thus we may apply Rolle to h, to deduce there is some point c with $a < c < b$ such that $h'(c) = 0$. But

$$h'(c) = f'(c)(g(b) - g(a)) - g'(c)(f(b) - f(a))$$

Thus

$$f'(c)(g(b) - g(a)) = g'(c)(f(b) - f(a))$$

This is one form of the Cauchy Mean Value Theorem for f and g. If $g'(c) \neq 0$ for any possible c, then the Mean Value theorem shows that $g(b) - g(a) \neq 0$, and so we can divide the above result to get

$$\frac{f(b)-f(a)}{g(b)-g(a)} = \frac{f'(c)}{g'(c)},$$

Note: Taking $g(x) = x$ recovers the Mean Value Theorem.

l'Hopital Revisited

We can get a much more useful form of l'Hopital's rule using the Cauchy Mean Value Theorem, directly from the definition of the derivative.

l'Hopital's Rule: General Form

Proposition: Let f and g be functions such that $f(a) = g(a) = 0$, and suppose that f and g are differentiable on an open interval I containing a, and that $g'(x) \neq 0$, except perhaps at a. Then

$$\lim_{x \to \alpha} \frac{f(x)}{g(x)} = \lim_{x \to \alpha} \frac{f'(x)}{g'(x)},$$

provided the second limit exists.

Proof: Pick $x > a$ and apply the Cauchy Mean Value Theorem to the interval $[a, x]$, to find c with $a < c < x$ such that

$$\frac{f(x)}{g(x)} = \frac{f(x) - f(a)}{g(x) - g(a)}$$

$$= \frac{f'(c)}{g'(c)}$$

Then $\lim_{x \to a+} \frac{f(x)}{g(x)} = \lim_{c \to a+} \frac{f'(c)}{g'(c)} = \lim_{x \to a+} \frac{f'(x)}{g'(x)}$, since we know the actual limit (not just the one sided limit) exists. Now repeat with $x < a$ to get the result.

Example: Evaluate $\lim_{x \to 0} \frac{1 - \cos x}{x^2} = \frac{1}{2}$.

Solution: We have

$$\lim_{x \to 0} \frac{1 - \cos x}{x^2} = \lim_{x \to 0} \frac{\sin x}{2x} = \frac{1}{2}$$

where the use of l'Hopital is justified since the second limit exists.

Note that you can't differentiate top and bottom again, and still expect to get the correct answer; one of the hypotheses of l'Hopital is that the first quotient is of the 0/0 form.

Example: Use l'Hopital to establish the following:

$$\lim_{x\to 0}\frac{\sqrt{1+x}-1-x/2}{x^2}=\frac{1}{8}$$

Solution: We have

$$\lim_{x\to 0}\frac{\sqrt{1+x}-1-x/2}{x^2}=\lim_{x\to 0}\frac{(1/2)(1+x)^{-1/2}-1/2}{2x}$$

$$=\lim_{x\to 0}\frac{(1/2)(-1/2)(1+x)^{-3/2}}{2}=\frac{-1}{8}$$

The use of l'Hopital is justified the second time, since the third limit exists; since we now know the second limit exists, the use of l'Hopital is justified the first time.

Infinite Limits

We can use Proposition to get results about infinite limits.

Example: Evaluate $\lim_{x\to\infty} x\log\left(1+\frac{1}{x}\right)$

Solution: $\lim_{x\to\infty} x\log\left(1+\frac{1}{x}\right)=\lim_{x\to\infty}\frac{\log(1+y)}{y}$ writing $y=1/x$,

$$=\lim_{y\to 0}\frac{\log(1+y)}{y}=\lim_{y\to 0}\frac{\log(1+y)}{y}=1$$

The last step is valid, since the final limit exists by l'Hopital; note also that this gives another way of finding $a_n = (1 + 1/n)^n$.

l'Hopital's Rule: Infinite Limits

Proposition: Let f and g be functions such that $\lim_{x\to\infty} f(x) = \lim_{x\to\infty} g(x) = \infty$, and suppose that $\lim_{x\to\infty} \frac{f'(x)}{g'(x)}$ exists. Then

$$\lim_{x\to\infty} \frac{f(x)}{g(x)} = \lim_{x\to\infty} \frac{f'(x)}{g'(x)}$$

Proof: Pick $\in > 0$ and choose a such that

$$\lim_{x\to\infty} \left|\frac{f'(x)}{g'(x)}\right| - l < \in \quad \text{for all } x > a.$$

Then pick K such that if $x > K$, then $g(x) - g(a) \neq 0$. By Cauchy,

$$\frac{f'(c)}{g'(c)} = \frac{f(x)-f(a)}{g(x)-g(a)} \quad \text{for all } x > K.$$

Note that although c depends on x, we always have $c > a$. Then

$$\frac{f(x)}{g(x)} = \frac{f(x)-f(a)}{g(x)-g(a)} \cdot \frac{f(x)}{f(x)-f(a)} \cdot \frac{g(x)-g(a)}{g(x)}$$

Taylor's Theorem

We have so far explored the Mean Value theorem, which can be rewritten as

$$f(a + h) = f(a) + hf'(c)$$

where c is some point between a and $a + h$. We have already met the approximation

$$f(a + h) \sim f(a) + hf'(a)$$

when we studied the Newton - Raphson method for solving an equation, and have already observed that the Mean Value

Theorem provides a more accurate version of this. Now consider what happens when f is a polynomial of degree n

$$f(x) = a_0 + a_1 x + a_2 x^2 + \ldots + a_{n-1} x^{n-1} + a_n x^n.$$

Note that $f(0) = a_0$. Differentiating gives

$$f'(x) = a_1 + 2a_2 x + 3a_3 x^2 \ldots + (n-1)a_{n-1} x^{n-2} + na_n x^{n-1},$$

and so $f'(0) = a_1$. Again, we have

$$f''(x) = 2a_2 + 3.2a_3 x \ldots + (n-1)(n-2)a_{n-1} x^{n-3} + n(n-1)a_n x^{n-2},$$

and $f¢¢(0) = 2a_2$. After the next differentiation, we get $f'''(0) = 3!a_3$, while after k differentiations, we get, $f^{(k)}(0) = k!a_k$, provided $k \leq n$. Thus we can rewrite the polynomial, using its value, and the value of its derivatives at 0, as

$$f(x) = f(0) + f'(0)x + \frac{f''(0)}{2!}x^2 + \frac{f'''(0)}{3!} x^3 + \ldots + \frac{f^{(n-1)}(0)}{(n-1)!} x^{n-1} + \frac{f^{(n)}(0)}{n!} x^n.$$

This opens up the possibility of representing more general functions than polynomials in this way, and so getting a generalisation of the Mean Value Theorem.

Taylors Theorem - Lagrange form of Remainder

Theorem: Let f be continuous on $[a, x]$, and assume that each of $f', f'', \ldots, f^{(n+1)}$ is defined on $[a, x]$. Then we can write

$$f(x) = P_n(x) + R_n(x),$$

where $P_n(x)$, the Taylor polynomial of degree n about a, and $R_n(x)$, the corresponding remainder, are given by

$$P_n(x) = f(a) + f'(a)(x-a) + \frac{f''(a)}{2!}(x-a)^2 + \frac{f^{(n)}(a)}{n!}(x-a)^n,$$

$$R_n(x) = \frac{f^{(n+1)}(c)}{(n+1)!}(x-a)^{n+1},$$

where c is some point between a and x. We make no attempt to prove this, although the proof can be done with the tools we have at our disposal. Some quick comments:

1. The theorem is also true for $x < a$; just restate it for the interval $[x, a]$ etc;
2. If $n = 0$, we have $f(x) = f(a) + (x - a)f'(c)$ for some c between a and x; this is a restatement of the Mean Value Theorem;
3. If $n = 1$, we have

$$f(x) = f(a) + (x - a)f'(x) + \frac{f''(c)}{2!}(x - a)^2$$

for some c between a and x; this often called the Second Mean Value Theorem;

4. In general we can restate Taylor's Theorem as

$$f(x) = f(a) + (x - a)f'(x) + \ldots + \frac{f^{(n)}(a)}{2!}(x - a)^n$$

$$+ \frac{f^{(n+1)}(c)}{(n+1)!}(x - a)^{n+1},$$

for some c between a and x;

5. The special case in which $a = 0$ has a special name; it is called Maclaurin's Theorem;
6. Just as with Rolle, or the Mean Value Theorem, there is no useful information about the point c.

We now explore the meaning and content of the theorem with a number of examples.

Example: Find the Taylor polynomial of order n about 0 for $f(x) = e^x$, and write down the corresponding remainder term.

Solution: There is no difficulty here in calculating derivatives clearly $f^{(k)}(x) = e^x$ for all k, and so $f^{(k)}(0) = 1$. Thus by Taylor's theorem,

$$e^x = 1 + x + \frac{x^2}{2!} + \frac{x^2}{2!} + \ldots \frac{x^n}{n!} + \frac{x^{n+1}}{(n+1)!}e^c$$

for some point c between 0 and x. In particular,

$$P_n(x) = 1 + x + \frac{x^2}{2!} + \frac{x^2}{2!} + \ldots \frac{x^n}{n!}$$

and $$R_n(x) = R_n(x) = \frac{x^{n+1}}{(n+1)!}e^c.$$

We can actually say a little more about this example if we recall that x is *fixed*. We have

$$e^x = P_n(x) + R_n(x) = P_n(x) + \frac{x^{n+1}}{(n+1)!}e^c$$

We show that $R_n(x) \to 0$ as $n \to \infty$, so that (again for fixed x), the sequence $P_n(x) \to e^x$ as n. If $x < 0$, $e^c < 1$, while if $x \geq 1$, then since $c < x$, we have $e^c < e^x$. thus

$R_n(x) = e\max(e, 1)0$ as n.

$$\left|R_{n(x)}\right| = \left|\frac{x^{n+1}}{(n+1)!}e^c\right| \leq \frac{|x|^{n+1}}{(n+1)!}$$

$\max(e^x, 1) \to 0$ as $n \to \infty$.

We think of the limit of the polynomial as forming a series, the Taylor series for e^x. We study series (and then Taylor series).

Example: Find the Taylor polynomial of order 1 about a for $f(x) = e^x$, and write down the corresponding remainder term.

Solution: Using the derivatives computed above, by Taylor's theorem,

$$e^x = e^\alpha + (x-a)e^\alpha + \frac{(x-a)^2}{2!}e^c$$

for some point c between a and x. In particular,

$$P_1(x) = e^{\alpha} + (x-a)e^{\alpha} \text{ and } R_1(x) = \frac{(x-a)^2}{2!}e^{c}.$$

Example: Find the Maclaurin polynomial of order $n > 3$ about 0 for $f(x) = (1 + x)^3$, and write down the corresponding remainder term.

Solution: We have

$$f(x) = (1 + x)^3$$
$$f'(x) = 3(1 + x)^2$$
$$f''(x) = 6(1 + x)$$
$$f'''(x) = 6$$
$$f^{(n)}(x) = 0 \text{ if } n > 3.$$

$$f(0) = 1$$

$$f'(x) = 3$$

$$f''(x) = 6$$

$$f'''(x) = 6$$

and so, by Taylor's theorem

$$(1 + x)^3 = 1 + 3x + \frac{6}{2!}x^2 + \frac{6}{3!}x^3,$$

a result we could have got directly, but which is at least reassuring.

Find the Taylor polynomial of order n about 0 for $f(x) = \sin x$, and write down the corresponding remainder term.

Solution: There is no difficulty here in calculating derivatives we have

$$f(x) = \sin x$$
$$f'(x) = \cos x$$
$$f''(x) = -\sin x$$
$$f'''(x) = -\cos x$$
$$f^{(4)}(x) = \sin x \text{ and so on.}$$

$$f(0) = 0$$

$$f'(x) = 1$$

$$f''(x) = 0$$

$$f'''(x) = -1$$

Thus by Taylor's theorem,

$$\sin x = x - \frac{x^3}{3!} + \frac{x^5}{5!} + \ldots + (-1)^{n+1} \frac{x^{2n+1}}{(2n+1)!}$$

Writing down the remainder term isn't particularly useful, but the important point is that

$$| R_{2n+1}(x) | \left| \frac{x^{2n+3}}{(2n+3)!} \right| \to 0 \text{ as } n \to \infty .$$

Exercise: Recall that $\cosh x = \dfrac{e^x + e^{-x}}{2}$, and that sinh,

$$x = \frac{e^x + e^{-x}}{2}.$$

Now check the shape of the following Taylor polynomials:

$$\cos x = 1 - \frac{x^2}{2!} + \frac{x^4}{4!} + \ldots + (-1)^n \frac{x^{2n}}{2n!} + \ldots$$

$$\sinh x = x + \frac{x^3}{3!} + \frac{x^5}{5!} + \ldots + \frac{x^{2n+1}}{(2n+1)!} + \ldots$$

$$\cosh x = 1 + \frac{x^2}{2!} + \frac{x^4}{4!} + \ldots + \frac{x^{2n}}{2n!} + \ldots$$

Example: Find the maximum error in the approximation

$$\sin(x) \sim x - \frac{x^3}{3!} \text{ given that } |x| < 1/2.$$

Solution: We use the Taylor polynomial for sin x of order 4 about 0, together with the corresponding remainder. Thus

$$\sin x = x - \frac{x^3}{3!} + \frac{x^5}{5!} \cos c$$

for some c with $0 < c < 1/2$ or $-1/2 < c < 0$. In any case, since $|x| < 1/2$,

$$\left|\frac{x^5}{5!}\cos c\right| \leq \left|\frac{x^5}{5!}\right| \leq \left|\frac{1}{2^5.5!}\right| \leq \frac{1}{120.32}$$

Warning: The Taylor polynomial always exists, providing f is suitably differentiable. But it need not be useful. Consider the example

$$(x) = \begin{cases} \exp\left(-1/x^2\right) \text{if } x > 0; \\ 0 \qquad \text{if } x \leq 0 \end{cases}$$

The interest in f is at 0; it is well behaved everywhere else. It turns out that,

$$f(0) = f¢(0) = f¢¢(0) = ... = f^{(n)}(0) = ... = 0.$$

So the Taylor polynomial of degree n for f about 0 is $P_n(x) = 0 + 0x + 0x^2 + ... + 0.x^n = 0$, and so for every n, $R_n(x) = f(x)$. Clearly in this case, P_n tells us nothing useful about the function.

Example: Find the Taylor polynomial of order n about 0 for $f(x) = (1 + x)^a$, and note that this gives a derivation of the binomial theorem. In fact, the remainder $|R_n(x)| \to 0$ as $n \to \infty$, provided $|x| < 1$.

Solution: There is again no difficulty here in calculating derivatives we have

$$f(x) = (1+x)^\alpha$$

$$f'(x) = \alpha(1+x)^{\alpha-1}$$

$$f''(x) = \alpha(\alpha-1)(1+x)^{\alpha-2}$$

$$f'''(x) = \alpha(\alpha-1)(\alpha-2)(1+x)^{\alpha-3}$$

and in general

$$f^{(n)}(x)=\alpha(\alpha-1)...(\alpha-n+1)(1+x)^{\alpha-n}$$

$$f(0)=1$$

$$f'(x)=\alpha$$

$$f''(x)=\alpha(\alpha-1)$$

$$f'''(x)=\alpha(\alpha-1)(\alpha-2)$$

$$f^{(n)}=\alpha(\alpha-1)...(\alpha-n+1)$$

$$f(x)=(1+x)^{\alpha}$$

$$f'(x)=\alpha(1+x)^{\alpha-1}$$

$$f''(x)=\alpha(\alpha-1)(1+x)^{\alpha-2}$$

$$f'''(x)=\alpha(\alpha-1)(\alpha-2)(1+x)^{\alpha-3}$$

and in General,

$$f^{(n)}(x)=\alpha(\alpha-1)...(\alpha-n+1)(1+x)^{\alpha-n}$$

$$f(0)=1$$

$$f'(x)=\alpha$$

$$f''(x)=\alpha(\alpha-1)$$

$$f'''(x)=\alpha(\alpha-1)(\alpha-2)$$

$$f^{(n)}=\alpha(\alpha-1)...(\alpha-n+1)$$

Thus by Taylor's theorem,

$$(1+x)^{\alpha}=1+\alpha x-\frac{\alpha(\alpha-1)}{2!}x^2+\frac{\alpha(\alpha-1)(\alpha-2)}{3!}x^3+$$

$$...+\frac{\alpha(\alpha-1)...(\alpha-n+1)}{n!}x^n+...$$

The remainder is not hard to deal with, but we omit the proof; in fact $|R_n(x)\to 0|$ when $n\to\infty$.

Note also that if $\alpha > 0$ is an *integer*, say $\alpha = n$ then $| R_n(x) | = 0$ and $f(x) = P_n(x)$. This is another way to get the Binomial theorem described.

Infinite Series

In this section we return to study a particular kind of sequence, those built by adding up more and more from a given collection of terms. One motivation comes from section, in which we obtained a sequence of approximating polynomials $\{P_n\}$ to a given function. It is more natural to think of adding additional terms to the polynomial, and as such we are studying series. However, there is a closely related sequence; the sequence of partial sums.

Arithmetic and Geometric Series Consider the Sum

$$a + (a + r) + (a + 2r) + \cdots + (a + nr)$$

We are trying to add up the terms in an arithmetic progression. A small amount of notation makes the addition easy.

Let $S_n = a + (a + r) + (a + 2r) + \cdots + (a + nr)$,

$S_n = (a + nr) + (a + (n - 1)r) + \cdots + (a + r) + a$,

so $2S_n = (2a + nr) + (2a + nr) + \cdots + (2a + nr)$, and

$$S_n = (n + 1).\frac{(\alpha + (\alpha + nr))}{2}.$$

Note that if $r > 0$ then $S_n \to \infty$ as $n \to \infty$, while if $r < 0$ then $S_n \to -\infty$ as $n \to \infty$.

We next consider a geometric progression (or series):

Let $S_n = a + ar + ar^2 + \cdots + ar^n$,

$rS_n = ar + ar^2 + ar^3 + \cdots + ar^{n+1}$,

so $(1 - r) S_n = a(1 - r^{n+1})$, and

$$S_n = \frac{a(1-r^{n+1})}{1-r} \quad \text{if} \quad r \neq 1.$$

Note that if $| r | < 1$, then $S_n \to \dfrac{a}{1-r}$ as $n \to \infty$.

Chapter 9

The Intermediate Value Theorem

We can draw the graph of a continuous function without lifting your pencil off the paper. This is made precise by the following result: Intermediate Value Theorem. Let $f(x)$ be a continuous function on the interval $[a, b]$. If $d \in [f(a), f(b)]$, then there is a $c \in [a, b]$ such that $f(c) = d$.

In the case where $f(a) > f(b)$, $[f(a), f(b)]$ is meant to be the same as $[f(b), f(a)]$. Another way to state the Intermediate Value Theorem is to say that the image of a closed interval under a continuous function is a closed interval.

We will present an outline of the proof of the Intermediate Value Theorem.

Example: Every polynomial of odd degree has at least one real root. We want to show that if $P(x) = a_n x^n + a_{n-1} x^{n-1} + \cdots + a_1 x + a_0$ is a polynomial with n odd and $a_n \neq 0$, then there is a real number c, such that $P(c) = 0$.

First let me remind you that it follows from the results in previous pages that every polynomial is continuous on the real line. There you also learned that

$$\lim_{x \to \infty} \frac{P(x)}{a_n x^n} = 1 \text{ and } \lim_{x \to -\infty} \frac{P(x)}{a_n x^n} = 1$$

Consequently for $|x|$ large enough, $P(x)$ and $a_n x^n$ have the same sign. But $a_n x^n$ has opposite signs for positive x and negative x.

Thus it follows that if $a_n > 0$, there are real numbers $x_0 < x_1$ such that $P(x_0) < 0$ and $P(x_1) > 0$.

Similarly if $a_n < 0$, we can find $x_0 < x_1$ such that $P(x_0) > 0$ and $P(x_1) < 0$. In either case, it now follows directly from the Intermediate Value Theorem that (for $d = 0$) there is a real number $c \in [x_0, x_1]$ with $P(c) = 0$.

The natural question arises whether every function which satisfies the conclusion of the Intermediate Value Theorem must be continuous. Unfortunately, the answer is no and counterexamples are quite messy. The easiest counterexample is the function

$$f(x) = \begin{cases} \sin\left(\dfrac{1}{x}\right) & ,\text{if } x \neq 0 \\ 0 & ,\text{if } x \neq 0 \end{cases}$$

This function fails to be continuous at $x = 0$. On the other hand, it is not too hard to see that $f(x)$ has the "Intermediate Value Property" even on closed intervals containing $x = 0$.

Intermediate Value Theorem

Theorem: Let f be continuous on the compact interval $[a, b]$, and assume that $f(a) < 0$ and $f(b) > 0$. Then there is some point c with $a < c < b$ such hat $f(c) = 0$.

Proof: We make no attempt to prove this formally, but sketch a proof with a pair of sequences and a repeated bisection argument. It is also noted that each hypothesis is necessary.

Example: Show there is at least one root of the equation $x - \cos x = 0$ in the interval [0, 1].

Proof: Apply the Intermediate Value Theorem to f on the closed interval [0, 1]. The function is continuous on that interval, and $f(0) = -1$, while $f(1) = 1 - \cos(1) > 0$. Thus there is some point $c \in (0, 1)$ such that $f(c) = 0$ as required.

Exercise: Show there is at least one root of the equation $x - e^{-x} = 0$ in the interval (0, 1).

Corollary: Let f be continuous on the compact interval $[a, b]$, and assume there is some constant h such that $f(a) < h$ and $f(b) > h$. Then there is a point c with $a < c < b$ such that $f(c) = h$.

Proof: Apply the Intermediate Value Theorem to $f - h$ on the closed interval [0, 1]. Note that by considering $-f + h$ we can cope with the case when $f(a) > h$ and $f(b) < h$.

Note: This theorem is the reason why continuity is often loosely described as a function you can draw without taking your pen from the paper. As we have seen with $y = 1/x$, this can give an inaccurate idea; it is in fact more akin to connectedness.

The Bisection Method

The Bisection Method at the same time gives a proof of the Intermediate Value Theorem and provides a practical method to find roots of equations. If your calculator can solve equations numerically, it most likely uses a combination of the Bisection Method and the Newton-Raphson Method.

Recall the statement of the Intermediate Value Theorem: Let $f(x)$ be a continuous function on the interval $[a, b]$. If $d \in [f(a), f(b)]$, then there is a $c \in [a, b]$ such that $f(c) = d$.

By replacing $f(x)$ by $f(x) - d$, we may assume that $d = 0$; it then suffices to obtain the following version: Let $f(x)$ be a continuous function on the interval $\in [a, b]$.

If $f(a)$ and $f(b)$ have opposite signs, then there is a $c \in [a, b]$ such that $f(c) = 0$. Here is an outline of its proof: Let's assume that $f(a) < 0$, while $f(b) > 0$, the other case being handled similarly. Set $a_0 = a$ and $b_0 = b$.

Now consider the midpoint,

$$m_0 = \frac{a_0 + b_0}{2},$$

and evaluate $f(m_0)$. If $f(m_0) < 0$, set $a_1 = m_0$ and $b_1 = b_0$. If $f(m_0) > 0$, set $a_1 = a_0$ and $b_1 = m_0$. (If $f(m_0)=0$, you're already done.) What situation are we in now? $f(a_1)$ and $f(b_1)$ still have opposite signs, but the length of the interval $[a_1, b_1]$ is only half of the length of the original interval $[a_0, b_0]$. Note also that $a_0 \leq a_1$ and that $b_0 \geq b_1$.

The second step: Consider the midpoint,

$$m_1 = \frac{a_1 + b_1}{2},$$

and evaluate $f(m_1)$. If $f(m_1) < 0$, set $a_2 = m_1$ and $b_2 = b_1$. If $f(m_1) > 0$, set $a_2 = a_1$ and $b_2 = m_1$. (If $f(m_1)=0$, you're already done.) What situation are we in now? $f(a_2)$ and $f(b_2)$ still have opposite signs, but the length of the interval $[a_2, b_2]$ is only a quarter of the length of the original interval $[a_0, b_0]$. Note also that $a_0 \leq a_1 \leq a_2$ and that

$$4b_0 \geq b_1 \geq b_2.$$

Continuing in this fashion we construct by induction two sequences,

$$(a_n)_{n=1}^{\infty} \text{ and } (b_n)_{n=1}^{\infty}$$

with the following properties:

1. (a_n) is an increasing sequence, (b_n) is a decreasing sequence.
2. $a_n \leq b_n$ for all n.
3. $f(a_n) < 0$ for all n, $f(b_n) > 0$ for all n.
4. $b_n - a_n = 2^{-n}(b - a)$ for all n.

It follows from the first two properties that the sequences (a_n) and (b_n) converge; set

$$\lim_{x\to\infty} a_n = a, \quad \lim_{x\to\infty} b_n = b.$$

The third property and the continuity of the function $f(x)$ imply that $f(a) \leq 0$ and that $f(b) \geq 0$. The crucial observation is the fact that the fourth property implies that $a = b$. Consequently, $f(a) = f(b) = 0$, and we are done.

The Derivative

The concept of Derivative is at the core of Calculus and modern mathematics. The definition of the derivative can be approached in two different ways. One is geometrical and the other one is physical.

Historically there was a fight between mathematicians which of the two illustrates the concept of the derivative best and which one is more useful.

We will not dwell on this and will introduce both concepts. Our emphasis will be on the use of the derivative as a tool.

Physical Concept of Derivative

This approach was used by Newton in the development of his Classical Mechanics. The main idea is the concept of velocity and speed. Indeed, assume you are traveling from point A to point B, what is the average velocity during the trip? It is given by

$$\text{Average velocity} = \frac{\text{distance from A to B}}{\text{time to get from A to B}}$$

If we now assume that A and B are very close to each other, we get close to what is called the instantaneous velocity. Of course, if A and B are close to each other, then the time it takes to travel from A to B will also be small. Indeed, assume that at time $t=a$, we are at A. If the time elapsed to get to B is Δt, then we will be at B at time $t = a + \Delta t$.

If Δs is the distance from A to B, then the average velocity is

$$\text{Average velocity } = \frac{\Delta s}{\Delta t}.$$

The instantaneous velocity (at A) will be found when Δt get smaller and smaller.

Here we naturally run into the concept of limit. Indeed, we have,

$$\text{Instantaneous velocity (at A)} = \lim_{\Delta t \to 0} \frac{\Delta s}{\Delta t}$$

If $f(t)$ describes the position at time t, then

$$\Delta s = f(a + \Delta t) - f(a).$$

In this case, we have

$$\text{Instantaneous velocity (at A)} = \lim_{\Delta t \to 0} \frac{f(a + \Delta t) - f(a)}{\Delta t}$$

Example: Consider a parabolic motion given by the function $f(t) = t^2$.

The instantaneous velocity at $t=a$ is given by,

$$\lim_{\Delta t \to 0} \frac{f(a+\Delta t)-f(a)}{\Delta t} = \lim_{\Delta t \to 0} \frac{f(a+\Delta t)^2 - a^2}{\Delta t}$$

Since,

$$\frac{(a+\Delta t)^2 - a^2}{\Delta t} = \frac{2a\Delta t + \Delta t^2}{\Delta t} = 2a + \Delta t$$

we conclude that the instantaneous velocity at $t = a$ is 2 a.

This concept of velocity may be extended to find the rate of change of any variable with respect to any other variable. For example, the volume of a gas depends on the temperature of the gas.

So in this case, the variables are V (for volume) as a function of T (the temperature). In general, if we have $y = f(x)$, then the average rate of change of y with respect to x from $x = a$ to $x = a + \Delta x$, where $\Delta x \neq 0$, is,

$$\text{Average Rate} = \frac{\Delta y}{\Delta x} = \frac{f(a+\Delta x)-f(a)}{\Delta x}$$

As before, the instantaneous rate of change of y with respect to x at $x = a$, is. Instantaneous Velocity,

$$(\text{at } x = a) = \lim_{\Delta x \to 0} \frac{\Delta y}{\Delta x} = \lim_{\Delta x \to 0} \frac{f(a+\Delta x)-f(a)}{\Delta x}$$

Notation. Now we get to the hardest part. Since we can not keep on writing "Instantaneous Velocity" while doing computations, we need to come up with a suitable notation for it. If we write dx for Δx small, then we can use the notation

$$\text{Instantaneous Velocity (at } x = a) = \frac{dy}{dx}(a)$$

This is the notation introduced by Leibniz.

Computation of Derivative

We have seen in the previous page how the derivative is defined: For a function $f(x)$, its derivative at $x = a$ is defined by,

$$f'(a) = \lim_{x \to a} \frac{f(x)-f(a)}{x-a}$$

$$\lim_{h \to o} \frac{f(a+h)-f(a)}{h}$$

Let us give some examples.

Example: Let us start with the function $f(x) = x^2$. We have

$$\frac{f(a+h)-f(a)}{h} = \frac{(a+h)^2 - a^2}{h}$$

$$= \frac{2ah + h^2}{h} = 2a + h.$$

So,

$$\lim_{h \to 0} \frac{f(a+h)-a}{h} = 2a$$

which means $f'(a) = 2a$.

What about the derivative of $f(x) = x^n$. Similar calculations, using the binomial expansion for $(x+y)^n$ (Pascal's Triangle), yield

$$f'(a) = na^{n-1}$$

Example: Let us consider the function,

$$f(x) = x\sin\left(\frac{1}{x}\right)$$

for $x \neq 0$, with $f(0) = 0$.

We have,

$$\frac{f(0+h)-f(0)}{h} = \frac{h\sin\left(\frac{1}{h}\right)}{h} = \sin\left(\frac{1}{h}\right)$$

Recall that the function $\sin\left(\frac{1}{x}\right)$ has no limit when x goes to 0. So the function has no derivative and no half-derivatives as well at x=0.

Example: Consider the function,

$$f(x)=\sqrt{|x|}\,.$$

Then we have,

$$\frac{f(0+h)-f(0)}{h}=\frac{\sqrt{|h|}}{h}$$

$$=\frac{1}{\sqrt{|h|}}$$

Since,

$$\lim_{h\to 0}\frac{1}{\sqrt{|h|}}\text{s}=+\infty$$

then f '(0) does not exist. But observe that the graph as a geometric figure has a tangent-albeit vertical: In fact, the way the concept of the tangent line was introduced is based on the notion of slope. You already know that vertical lines do not have slopes.

It can be quite laborious (or impossible) to compute the derivative by hand as we have done so far. In the next pages we will show how techniques of differentiation help bypass the limit calculations and make our life much easier.

Example: Consider the function $f(x)=1/x$ for $x\neq 0$. We have

$$\frac{f(a+h)-f(a)}{h}=\frac{\frac{1}{a+h}-\frac{1}{a}}{h}$$

$$=\frac{a-(a+h)}{ha(a+h)}$$

$$=\frac{-h}{ha(a+h)}$$

Consequently,

$$\lim_{h\to 0}\frac{f(a+h)-f(a)}{h}=\lim_{h\to 0}\frac{-1}{a(a+h)}=-\frac{1}{a^2}.$$

The algebraic trick in both of the examples above has been to factor out "h" in the numerator, so that we can cancel it with the "h" in the denominator! This is what you try to do whenever you are asked to compute a derivative using the limit definition.

Example: Let us discuss the derivative of $f(x) = |x|$ at 0. We have

$$\frac{f(0+h)-f(a)}{h}=\frac{|h|-0}{h}=\frac{|h|}{h}$$

But

$$\lim_{h\to 0+}\frac{|h|}{h}=1 \text{ and } \lim_{h\to 0-}\frac{|h|}{h}=-1$$

which implies that $f'(0)$ does not exist.

Remark. This example is interesting. Even though the derivative at the point does not exist, the right and the left limit of the ratio do exist. In fact, if we use the slope-interpretation of the derivative we see that this means that the graph has two lines close to it at the point under consideration. They could be seen as "half-tangents". See Picture.

Techniques of Differentiation

Maybe the easiest and most useful formulas are the ones that say that the derivative is linear:

$$(f+g)'(a)=f'(a)+g'(a)$$

$$(cf)'(a)=cf'(a)$$

Combined with the formula $(x^n)' = n\, x^{n-1}$, we see that every polynomial function has a derivative at any point.

Example: For $P(x) = 1-2x + 3x^4 -5\,x^6$, we have

$$P'(x) = -2+12x^3\text{-}30x^5$$

The next two formulas are the most powerful ones. They deal with the derivative of a product and a quotient. They are commonly called the product rule and the quotient rule. We have

$$(fg)'(a) = f'(a)g(a) + f(a)g'(a)$$

$$\left(\frac{f}{a}\right)'(a) = \frac{f'(a)g(a) - f(a)g'(a)}{g^2(a)}$$

In particular, we have

$$\left(\frac{1}{f}\right)'(a) = -\frac{f'(a)}{f^2(a)}$$

So, we have

$$\left(\frac{1}{x^n}\right) = -\frac{nx^{n-1}}{x^{2n}} = -nx^{-n-1}$$

which means that the formula $(x^r)' = r\,x^{r-1}$ is also valid for negative exponents. Before we discuss the derivative of trigonometric functions, let us stop here and reflect a little bit more on polynomial functions. Indeed, we saw that the derivative of a polynomial function is also a polynomial function.

So we can take another derivative and generate a new function. This function is called the second derivative. We can keep doing this as long as we want to. The functions obtained are called higher derivatives. The common notations used for them are

$$f'(x),\quad f''(x),\quad f'''(x),\quad f^{(4)}(x),\quad \ldots,\quad f^{(n)}(x)$$

$$y',\quad y'',\quad y''',\quad y^{(4)},\quad \ldots,\quad y^{(n)}$$

$$\frac{dy}{dx},\quad \frac{d^2y}{dx^2},\quad \frac{d^3y}{dx^3},\quad \frac{d^4y}{dx^2}\quad \ldots,\quad \frac{d^ny}{dx^n}$$

The Derivatives of Trigonometric Functions

$$\sin' x = \cos x \qquad \cos' x = -\sin x$$

$$\tan' x = \sec^2 x \qquad \cot' x = -\csc^2 x$$

$$\sec' x = \sec x \tan x \quad \csc' x = -\csc x \cot x$$

Trigonometric functions are useful in our practical lives in diverse areas such as astronomy, physics, surveying, carpentry etc.

Using the derivative language, this limit means that $\sin'(0) = 1$. This limit may also be used to give a related one which is of equal importance:

$$\lim_{x \to 0} \frac{\cos(x) - 1}{x} = 1$$

To see why, it is enough to rewrite the expression involving the cosine as,

$$\frac{\cos(x) - 1}{x} = \frac{(\cos(x) - 1)(\cos(x) + 1)}{x(\cos(x) + 1)}$$

$$= \frac{(\cos^2(x) - 1)}{x(\cos(x) + 1)}$$

But,

$$\cos^2(x) - 1 = -\sin^2(x),$$

so we have

$$\lim_{x \to 0} \frac{\cos(x) - 1}{x} = \lim_{x \to 0} \frac{-\sin^2(x)}{x(\cos(x) + 1)}$$

$$= \lim_{x \to 0} x \frac{-\sin^2(x)}{x^2(\cos(x) + 1)} = 0$$

In fact, we may use these limits to find the derivative sin (x) of and sin (x) at any point $x = a$. Indeed, using the addition formula for the sine function, we have

$$\sin(a+h) = \sin(a)\cos(h)+\sin(h)\cos(a)$$

So,

$$\frac{\sin(a+h)}{h} = \frac{\sin(a)\cos(h)-1}{h}+\cos(a)\frac{\sin(h)}{h}$$

which implies,

$$\lim_{h\to 0}\frac{\sin(a+h)-\sin(a)}{h} = \cos(a)$$

So we have proved that sin'(a) exists and sin'(a) = cos(a). Similarly, we obtain that cos' (a) exists and that cos'(a) = – sin(a). Since tan(x), cot(x), sec (x), and csc(x)are all quotients of the functions and, we can compute their derivatives with the help of the quotient rule:

$$\frac{d}{dx}(\tan(x)) = \sec^2(x) = 1+\tan^2(x)$$

$$\frac{d}{dx}(\sec(x)) = \sec(x) = 1+\tan(x)$$

$$\frac{d}{dx}(\tan(x)) = -\csc^2(x) = -1-\cot^2(x)$$

$$\frac{d}{dx}(\csc(x)) = -\csc(x)\cot(x)$$

It is quite interesting to see the close relationship between tan (x) and (and also between cot (x) and csc (x)).

From the above results we get

$$\sin''(x) = -\sin(x)\text{and}\cos'' = -\cos(x).$$

These two results are very useful in solving some differential equations.

Example: Let $f(x) = \sin(2x)$. Using the double angle formula for the sine function, we can rewrite

$$\sin(2x) = 2\sin(x)\cos(x)$$

So using the product rule, we get

$$\frac{d}{dx}(\sin(2x)) = 2(\cos(x)\cos(x) - \sin(x)\sin(x))$$

$$= 2\left(\cos^2(x) - \sin^2(x)\right)$$

which implies, using trigonometric identities,

$$\frac{d}{dx}(\sin(2x)) = 2\cos(2x)$$

In fact next we will discuss a formula which gives the above conclusion in an easier way.

Implicit Differentiation

In many examples, especially the ones derived from differential equations, the variables involved are not linked to each other in an explicit way. Most of the time, they are linked through an implicit formula, like $F(x,y) = 0$. Once x is fixed, we may find y through numerical computations. (By some fancy theorems, we may formally show that y may indeed be seen as a function of x over a certain interval). The question becomes what is the derivative $\frac{dy}{dx}$, at least at a certain a point? The method of implicit differentiation answers this concern. Let us illustrate this through the following example.

Example: Find the equation of the tangent line to the ellipse

$$25\,x^2 + y^2 = 109$$

at the point (2,3). One way is to find y as a function of x from the above equation, then differentiate to find the slope of the tangent line.

We will leave it to the reader to do the details of the calculations. Here, we will use a different method. In the above equation, consider y as a function of x:

$$25\,x^2 + y(x)^2 = 109,$$

and use the techniques of differentiation, to get

$$\frac{d}{dx}\left[25x^2 + y(x)^2\right] = 50x + 2y\frac{dy}{dx} = 0$$

From this, we obtain

$$\frac{dy}{dx} = -25\frac{x}{y}$$

which implies that $\frac{dy}{dx} = -\frac{50}{3}$ at the point (2,3). So the equation of the tangent line is given by,

$$y - 3 = -\frac{50}{3}(x-2),$$

or equivalently,

$$3y + 50\,x = 109.$$

You may wonder why bother if this is just a different way of finding the derivative? Consider the following example! It can be very hard or in fact impossible to solve explicitly for y as a function of x.

Example: Find y' if

$$y^4 + 4y - 3x^3\sin(y) = 2x + 1$$

This is a wonderful example of an implicit relation between $x$and y. So how do we find y'?

Let us differentiate the above equation with respect to x where y is considered to be a function of x. We get

$$4y^3y' + 4y' - 9x^2\sin(y) - 3x^3\cos(y)y' = 2$$

Easy algebraic manipulations give

$$y' = \frac{2 + 9x^2\sin(y)}{4y^3 + 4 - 3x^3\cos(y)}$$

We can also find higher derivatives of y such as y'' in this manner. We only have to differentiate the above result. Of course the calculations get little more messy.

The Chain Rule

$$(f\,o\,g(x))' = g'(x)\,f'(g(x))$$

or
$$\frac{dy}{dx} = \frac{dy}{du}\frac{du}{dx}$$

As a motivation for the chain rule, consider the function

$$f(x) = (1+x^2)^{10}.$$

Since $f(x)$ is a polynomial function, we know from previous pages that $f'(x)$ exists.

Naturally one may ask for an explicit formula for it. One tedious way to do this is to develop $(1+x^2)^{10}$ using the Binomial Formula and then take the derivative. Of course, it is possible to do this, but it won't be much fun. But what if we have to deal with $(1+x^2)^{100}$! Then I hope you agree that the Binomial Formula is not the way to go anymore. So what do we do? The answer is given by the Chain Rule.

Before we discuss the Chain Rule formula, let us give another example.

Example: Let us find the derivative of $f(x) = \sin(2x)$. One way to do that is through some trigonometric identities. Indeed, we have

$$\sin(2x) = 2\sin(x)\cos(x)$$

So we will use the product formula to get

$$(\sin(2x))' = 2(\sin'(x)\cos(x) + \sin'(x)\cos'(x))$$

which implies,

$$(\sin(2x))' = 2(\cos^2(x) - \sin^2(x))$$

Using the trigonometric formula

$$\cos(2x) = \cos^2(x) - \sin^2(x),$$ we get

$$(\sin(2x))' = 2\cos(2x)$$

Once this is done, you may ask about the derivative of sin $(5x)$? The answer can be found using similar trigonometric identities, but the calculations are not as easy as before. Again we will see how the Chain Rule formula will answer this question in an elegant way. In both examples, the function $f(x)$ may be viewed as:

$$f(x) = h(g(x))$$

where $g(x) = 1+x^2$ and $h(x) = x^{10}$ in the first example, and $h(x) = \sin(x)$and $g(x) = 2x$ in the second. We say that $f(x)$ is the composition of the functions $g(x)$ and $h(x)$ and write

$$f(x) = h\text{o}g(x)$$

The derivative of the composition is given by the formula

$$(f\text{o}g(x))' = g'(x)f'(g(x))$$

Another way to write this formula is,

$$\frac{dy}{dx} = \frac{dy}{du}\frac{du}{dx}$$

where,

$$y = f\text{o}g(x) = f(u)$$

and $u = g(x)$. This second formulation (due to Leibniz) is easier to remember and is the formulation used almost exclusively by physicists.

Example: Let us find the derivative of

$$f(x) = (1+x^2)^{100}$$

We have,

$$f(x) = h\text{o}g(x),$$

where,

$$g(x) = 1+x^2 \text{ and } h(x) = x^{100}.$$

Then the Chain rule implies that $f'(x)$ exists, which we knew since it is a polynomial function, and,

$$f'x = 2x\cdot\left[100(1+x^2)^{99}\right]$$

$$= 200x\left(1+x^2\right)^{99}$$

Example: Let us find the derivative of

$$f(x) = \sin(5x)$$

We have $f(x) = h\text{o}g(x)$, where

$$g(x) = 5x \text{ and}$$

$$h(x) = \sin\ (5x)\ .$$

Then the Chain rule implies that $f\,'(x)$ exists and

$$f\,'(x) = 5\cdot[\cos(5x)]$$

$$= 5\cos(5x)$$

In fact, this is a particular case of the following formula

$$[f(ax+b)]' = af\,'(ax+b)$$

The following formulas come in handy in many areas of techniques of integration.

$$\frac{d}{dx}\left(u^n\right) = nu^{n-1}\frac{du}{dx}$$

$$\frac{d}{dx}(\cos(u)) = -\sin(u)\frac{du}{dx}$$

$$\frac{d}{dx}(\sin(u)) = \cos(u)\frac{du}{dx}$$

$$\frac{d}{dx}(\text{In}(u)) = \frac{1}{u}\frac{du}{dx}$$

$$\frac{d}{dx}\left(d^u\right) = e^u\frac{du}{dx}$$

More formulas for derivatives can be found in our section of tables.

Chapter 10

Differentiating Functions

Differentiating Sums of Functions

$f(x)$ equals $x^{n_1} + x^{n_2}$ which is a sum of two functions of x, $A(x)$ and $b(x)$ where $A(x) = x^{n1}$ and $B(x) = x^{n2}$ Therefore, if $f(x) = A(x) + B(x)$ What would $f'(x)$be?

The answer is that the derivative is the sum of the derivatives of the two functions To prove this let us return to the definition of the derivative.

We can express a small change in f, Δf, equal to $f(x+\Delta x) - f(x)$.

Therefore:

$$\Delta f = f(x+\Delta x) - f(x) = [A(x+\Delta x) + B(x+\Delta x)] \\ -[A(x) + B(x)]$$

Δx and taking the limit as Δx goes to zero gives us the instantaneous rate of change of f with respect to x.

$$\frac{df}{dx} = \lim_{\Delta x \to 0} \frac{f(x+\Delta x) - f(x)}{\Delta x}$$

$$= \frac{[A(x+\Delta x) + B(x+\Delta x)] - [A(x) + B(x)]}{\Delta x}$$

Combing the $A(x)$ and $B(x)$ terms together simplifies the above expression to:

$$\frac{df}{dx} = \lim_{\Delta x \to 0} \frac{f(x+\Delta x) - f(x)}{\Delta x}$$

$$= \lim_{\Delta x \to 0} \frac{[A(x+\Delta x) - A(x)]}{\Delta x} + \frac{[B(x+\Delta x) - B(x)]}{\Delta x}$$

Which reduces to:

$$\frac{df}{dx} = f'(x) = A'(x) + B'(x)$$

Therefore if $f(x)$ is a sum of two functions of x, then its derivative *with respect to* x is the sum of the derivatives of the functions with respect to x. Thus:

$$f(x) = x^{n_1} + x^{n_2}$$

$$\frac{df}{dx} = f'(x) = n_1 \cdot x^{n_1 - 1} + n_2 \cdot x^{n_2 - 1}$$

Similarly if $f(x)$ is defined in terms of a difference among some functions of x, then $f'(x)$ is the sum of the difference among the derivatives of the functions.

$$f(x) = x^{n_1} - x^{n_2} - x^{n_3} \ldots\ldots\ldots$$

$$\frac{df}{dx} = f'(x) = n_1 \cdot x^{n_1 - 1} - n_2 \cdot x^{n_2 - 1} - n_3 \cdot x^{n3 - 1} \ldots\ldots.$$

Differentiating Products of Functions

Consider the following function:

$$f(x) = x^2 \cdot (x+1)^3$$

If we let , then $f(x)$ can be expressed as the product of the two function $A(x)$ and $B(x)$ such that:

$$f(x) = A(x) \cdot B(x)$$

We can differentiate products of functions by using the definition of the derivative. A small change in f can be written as:

$$\Delta f = f(x+\Delta x) - f(x) = A(x+\Delta x)B(x+\Delta x) - A(x)B(x)$$

Next, divide by Δx to calculate the rate of change of f with respect to f:

$$\frac{\Delta f}{\Delta x} = \frac{f(x+\Delta x) - f(x)}{\Delta x}$$
$$= \frac{A(x+\Delta x)B(x+\Delta x) - A(x)B(x)}{\Delta x}$$

Taking the limit as Δx goes to zero gives us the instantaneous rate of change of f with respect to x, or the derivative of $f(x)$.

$$\frac{df}{dx} = \lim_{\Delta x \to 0} \frac{f(x+\Delta x) - f(x)}{\Delta x}$$
$$= \frac{A(x+\Delta x)B(x+\Delta x) - A(x)B(x)}{\Delta x}$$

From the definition of the derivative we know that:

$$\lim_{\Delta x \to 0} \frac{f(x+\Delta x) - f(x)}{\Delta x} = f'(x)$$

Multiplying both sides by this infinitely small Δx

$$f(x+\Delta x) - f(x) = f'(x) \cdot \Delta x$$
$$f(x+\Delta x) = f(x) + f'(x).\Delta x$$

Since both $A(x)$ and $B(x)$ are functions of x, then $A(x+\Delta x)$ and $B(x+\Delta x)$ can be substituted with $A(x) + A'(x) \cdot \Delta x$ and $B(x) + B'(x) \cdot \Delta x$ respectively. Note that this substitution only holds true for Δx going to zero. We now have:

Canceling terms and dividing through by Δx reduces it to:

$$f'(x) = \lim_{\Delta x \to 0} A(x)B'(x) + B(x)A'(x) + A'(x)B'(x)\Delta x$$

$$\therefore \qquad f'(x) = A(x)B'(x) + B(x)A'(x)$$

Thus the derivative of a function $f(x)$ that is a product of two functions of x, is simply the product of the first function and the derivative of the second function plus the product of the second function and the derivative of the first function.

$$f(x) = A(x) \cdot B(x)$$
$$f'(x) = A(x)B'(x) + B(x)A'(x)$$

Differentiating Functions of any Power N

We have proven that if $f(x) = x^n$ then $f'(x) = n \cdot x^{n-1}$ for n equal to a positive integer, i.e. 1, 2, 3, 4 etc. What if n were a fraction such that $f(x) = x^{\frac{1}{3}}$.

As we shall prove, the derivative of any function of x of the form $f(x) = x^n$ is always $f'(x) = n \cdot x^{n-1}$ where n is any real number, positive, negative, or fraction. This is no coincidence but is because of the way exponents are defined as a continuous operation for any n.

Let us consider the first case of $f(x) = x^{\frac{1}{n}}$, where n is any positive integer and 1/n is a fraction.

$$f(x) = x^{\frac{1}{n}}$$
$$f = x^{\frac{1}{n}}$$

Raising both sides to the *n'th* power:

$$f^n = x$$
$$x = f^n$$

Differentiating *x*, with respect to *f* yields:

$$f^n = x$$
$$x = f^n$$

Taking the reciprocal of the function:

$$\frac{dx}{df} = \frac{1}{n \cdot f^{n-1}} = \frac{1}{n} \cdot \frac{1}{f^{(n-1)}}$$

From the definition of the function, we know that $f = x^{\frac{1}{n}}$. Making this substitution:

$$\frac{df}{dx} = \frac{1}{n} \cdot \frac{1}{x^{\frac{1}{n}\cdot(n-1)}}$$

$$\frac{df}{dx} = \frac{1}{n} \cdot \frac{1}{x^{\left(1-\frac{1}{n}\right)}}$$

$$\frac{df}{dx} = \frac{1}{n} \cdot x^{-\left[\left(1-\frac{1}{n}\right)\right]}$$

$$\therefore \qquad \frac{df}{dx} = f'(x) = \frac{1}{n} \cdot x^{\frac{1}{n}-1}$$

This concludes the proof that if $f(x) = x^n$ then $f'(x) = n \cdot x^{n-1}$, for any positive n, integer or fraction. We will prove it also holds true for n as a negative number. We can use the product rule to prove that the derivative of $f(x) = x^n$ is $f'(x) = n \cdot x^{n-1}$ for all n, negative real numbers also. Consider the function:

$$f(x) = x^{-n}$$

$$f = \frac{1}{x^n}$$

Multiplying both sides by x^n yields:

$$f.x^n = 1$$

$$f(x) \cdot x^n = 1$$

$$f(x) \cdot A(x) = B(x)$$

where $A(x) = x^n$ and $B(x) = 1$

The left side of the equation represents a product of two functions, *f* and x^n, and the right side is a constant function, $B(x) = c = 1$.

Since both functions are equal to each then their derivatives must be equal. Using the product rule to differentiate the right side with respect to *x* results in:

$$\frac{d[f(x)\cdot A(x)]}{dx} = f(x)\cdot A'(x) + A(x)\cdot f'(x)$$

Similarly, differentiating the right side with respect to *x* yields:

$$B(x) = 1\cdot x^0$$

$$B'(x) = 0$$

Setting the derivatives of the left and right side equal to each other:

$$f(x)\cdot A'(x) + A(x)\cdot f'(x) = 0$$

Remember that:

$$f(x) = x^{-n} = \frac{1}{x^n}$$

$$A(x) = x^n$$

$$A'(x) = n\cdot x^{n-1}$$

Making the required substitutions:

$$\left(\frac{1}{x^n}\right)\cdot nx^{n-1} + x^n.f'(x) = 0$$

$$\frac{n}{x} + x^n\cdot f'(x) = 0$$

We can now solve for $f'(x)$

$$f'(x) = -\left(\frac{n}{x}\right)\cdot\frac{1}{x^n}$$

$$f'(x) = -n\cdot\frac{1}{x^{1+n}}$$

$$f'(x) = -n\cdot x^{-(1+n)}$$

$$f'(x) = -n\cdot x^{-n+1}$$

Therefore if $f(x) = x^{-n}$ then $f'(x) = -n \cdot x^{-n+1}$, where –*n* is a negative integer.

We have proven that if $f(x) = x^n$ is $f'(x) = n \cdot x^{n-1}$ for positive real numbers and negative integers. What remains is to prove it true for negative fractional powers as well. To do this let,

$$f(x) = x^{-\frac{1}{n}}$$

$$f(x) = \frac{1}{x^{\frac{1}{n}}}$$

$$f(x) = x^{\frac{1}{n}} \cdot$$

Differentiating both sides with respect to *x* yields:

$$f(x) \cdot \frac{1}{n} . x^{\frac{1}{n}-1} + x^{\frac{1}{n}} \cdot f'(x) = 0$$

Substituting known values for *f(x)* and solving $f'(x)$ for gives us:

$$f'(x) = -\frac{1}{n} \cdot x^{-\frac{1}{n}-1}$$

Thus if $f(x) = x^{-n}$ then $f'(x) = -n \cdot x^{-n+1}$ for any *n*, positive or negative real number.

This is true because of the way exponents are defined as a uniform operation for any *n*.

The purpose of going through all the proofs for the different cases of *n* was to give you a better understanding of how to differentiate functions of *x* with respect to *x*.

Differentiating Functions of Functions

The last technique of differentiation is for differentiating functions of functions of *x* or functions of the form, $f(g(x)) = g(x)^n$.

For example consider the function: $f(x) = \left(x^2+1\right)^{\frac{1}{2}}$

The derivative is not $f'(x) = \frac{1}{2}\left(x^2+1\right)^{\frac{1}{2}-1}$ because we also need to take into consideration the inside function of *x*, x^2+1. We can replace x^2+1 with *g*(*x*) and get:

$$f(g(x)) = g(x)^{\frac{1}{2}}$$

$$f(g) = g^{\frac{1}{2}}$$

To find the derivative of $f(g(x))$ with respect to *x*, we first need to find the derivative of *f* with respect to *g*. From the definition of the derivative:

$$\frac{df}{dg} = f'(g) = \lim_{\Delta g \to 0} \frac{f(g+\Delta g)-f(g)}{\Delta g}$$

Next we find the derivative of *g*(*x*), the inside function, with respect to x.

$$\frac{dg}{dx} = g'(x) = \lim_{\Delta x \to 0} \frac{g(x+\Delta x)-g(x)}{\Delta x}$$

Now we multiply the two derivatives $f'(g)$ and $g'(x)$ to get df/dx:

$$f'(g)\cdot g'(x) = \frac{df}{dg}\cdot\frac{dg}{dx}$$

$$= \lim_{\Delta x \to 0} \frac{f(g+\Delta g)-f(g)}{\Delta x}\cdot\frac{g(x+\Delta x)-g(x)}{\Delta x}$$

$$\Delta g = g(x+\Delta x)-g(x) \quad f'(g)g'(x)$$

$$= \frac{df}{dg}\cdot\frac{dg}{dx} = \lim_{\Delta x \to 0} \frac{f(g+\Delta g)-f(g)}{g(x+\Delta x)-g(x)}\cdot\frac{g(x+\Delta x)-g(x)}{\Delta x}$$

Δx goes to zero, $\dfrac{g(x+\Delta x)-g(x)}{\Delta x}g'(x)$ or:

$$\frac{dg}{dx}=g'(x)$$

Similarly, Δg also goes to zero as Δx goes to zero. Multiplying both sides by *dx*.

$$dg=g'(x)\cdot dx$$

Thus as *dx* goes to zero;

$$dg=c.0=0$$

As *dx* goes to zero, $\dfrac{f(g+\Delta g)-f(g)}{\Delta g}$ becomes $f'(g)$.

At the same time the in the denominator of,

$$\frac{f(g+\Delta g)-f(g)}{\Delta g}$$

cancels out with Δg the in the numerator of ,

$$\frac{g(x+\Delta x)-g(x)}{\Delta x},$$

since they are both the equivalent. To conclude

$$\frac{df}{dg}\cdot\frac{dg}{dx}=\frac{df}{dx}=f'(g)g'(x)$$

Thus the derivative of a function of a function of x with respect to x, is the derivative of the outer function with respect to the inner function, multiplied by the derivative of the inner function with respect to *x*. Some examples will show how this is done.

Using the First Derivative

From the definition of the derivative, $f'(x)$ is the derivative of a function $f(x)$. The derivative of a function tells us how fast f is changing relative to the independent variable,

x. Thus the derivative refers specifically the rate of change of the anti-derivative function with respect to x. Rate of change is also synonymous with the slope of the tangent line to the graph at a particular point.

Therefore, a function and its derivative are closely related and knowing just the derivative can tell us a great deal about the behavior of its anti-derivative.

Since the derivative of a function is derived from the definition of a derivative as:

$$\lim_{\Delta \to 0} \frac{f(x+\Delta x)-f(x)}{\Delta x},$$

we can work with the definition to find the anti-derivative when only the derivative is known. Also recall how differentiation is based on a limiting or subtracting process and dividing; therefore working backwards would tell us that we should be adding and multiplying.

We will develop on this later, however, let us first look at how we can use $f'(x)$ to obtain equilibrium points on the graph of the anti-derivative or $f(x)$. Equilibrium points are by definition, points on the graph refer to static situations where the rate of change is zero.

Thus a change in the independent variable results in no change in the dependent variable. Equilibrium sate generally occurs when a situation has reached a critical maximum value and then decreases or where a situation has reached a critical minimum value then increases.

A change in concavity occurs at points on the graph called *inflection* points.

As we shall soon study, the derivatives value at an inflection does not have to be zero. Therefore we will restrict our definition of equilibrium points to reflect either maximum or minimum values on the graph.

For example to find equilibrium points for the function, $f(x)=x^3-x^2$. we first need to differentiate it to get, $f'(x)$. The derivative, $f'(x)$, tells us the instantaneous rate of change of function, $\frac{df}{dx}$, at any point x. Since we want to find points where the rate of change of $f'(x)$ is zero, we need to set equal to zero to find those values of x which the satisfy the equation, $f'(x) = 0$. Doing this for $f(x) = x^3 - x^2$ results in:

$$f'(x) = 3x^2 - 2x$$
$$f'(x) = 0$$
$$3x^2 - 2x = 0$$
$$(x)(3x - 2) = 0$$

$\therefore x = 0$ and $x = 2/3$

Chapter 11

Limit of a Function

x	$\frac{\sin x}{x}$
1	0.841471
0.1	0.998334
0.01	0.999983

Although the function (sin x)/x is not defined at zero, as x becomes closer and closer to zero, (sin x)/x becomes arbitrarily close to 1. In other words, the limit of (sin x)/x as x approaches zero equals 1.

In mathematics, the limit of a function is a fundamental concept in calculus and analysis concerning the behaviour of that function near a particular input. Formal definitions, first devised in the early 19th century, are given below. Informally, a function f assigns an output $f(x)$ to every input x. The function has a limit L at an input p if $f(x)$ is "close" to L whenever x is "close" to p. In other words, $f(x)$ becomes closer and closer to L as x moves closer and closer to p. More specifically, when f is applied to each input *sufficiently* close to p, the result is an output value that is *arbitrarily* close to L. If the inputs "close" to p are taken to values that are very different, the limit is said to *not exist.*

The notion of a limit has many applications in modern calculus. In particular, the many definitions of continuity employ the limit: roughly, a function is continuous if all of its limits agree with the values of the function. It also appears in the definition of the derivative: in the calculus of one variable, this is the limiting value of the slope of secant lines to the graph of a function.

History

Although implicit in the development of calculus of the 17th and 18th centuries, the modern idea of the limit of a function goes back to Bolzano who, in 1817, introduced the basics of the epsilon-delta technique to define continuous functions. However, his work was not known during his lifetime (Felscher 2000). Cauchy discussed limits in his *Cours d'analyse* (1821) and gave essentially the modern definition, but this is not often recognized because he only gave a verbal definition (Grabiner 1983). Weierstrass first introduced the epsilon-delta definition of limit in the form it is usually written today. He also introduced the notations lim and $\lim_{x \to x0}$ (Burton 1997). The modern notation of placing the arrow below the limit symbol is due to Hardy in his book *A Course of Pure Mathematics* in 1908 (Miller 2004).

Motivation

Imagine a person walking over a landscape represented by the graph of $y = f(x)$. His horizontal position is measured by the value of x, much like the position given by a map of the land or by a global positioning system. His altitude is given by the coordinate y. He is walking towards the horizontal position given by $x = p$. As he gets closer and closer to it, he notices that his altitude approaches L. Say there's a wall there so he can't stand on that point exactly, but can still get arbitrarily close to it. If asked about the altitude of $x = p$, he would then answer L. What, then, does it mean to say that his altitude approaches L? It means that his altitude gets nearer and nearer to L except for a possible small error in accuracy. For example, suppose a particular accuracy goal is set for our traveller: he must get within ten meters of L in altitude. He reports back that indeed he can get within ten meters of L, since he notes that when he is anywhere within fifty horizontal meters of p, his altitude is *always* ten meters or less from L. The accuracy goal is then changed: can he get within one vertical meter? Yes. If he is anywhere within seven horizontal meters of p, then his altitude always remains within one meter from the target L. In summary, to say that the traveller's altitude approaches L as his horizontal position approaches p means that for every target accuracy goal, however small it may be, there is some neighbourhood of p whose altitude fulfills that accuracy goal.

The initial informal statement can now be explicated: The limit of a function $f(x)$ as x approaches p is a number L with the following property: given any target distance from L, there is a distance from p within which the values of $f(x)$ remain within the target distance.

This explicit statement is quite close to the formal definition of the limit of a function with values in a topological space.

Example of a Function Without a Limit

Functions on Metric Spaces

Suppose M and N are subsets of metric spaces A and B, respectively, and $f : M \to N$ is defined between M and N, with $x \in M$, p a limit point of M and $L \in N$. It is said that the limit of f as x approaches p is L and write

$$\lim_{x \to p} f(x) = L$$

if the following property holds:

- For every e > 0, there exists a d > 0 such that $d_B(f(x), L) < \varepsilon$ whenever $0 < d_A(x, p) < \delta$.

Again, note that p need not be in the domain of f, nor does L need to be in the range of f, and even if $f(p)$ is defined it need not be equal to L.

An alternative definition using the concept of neighbourhood is as follows:

$$\lim_{x \to p} f(x) = L$$

if, for every neighbourhood V of L in B, there exists a neighbourhood U of p in A such that $f(U)\text{-}M - \{p\}) \subseteq V$.

Functions on topological spaces

Suppose X, Y are topological spaces with Y a Hausdorff space. Let p be a limit point of $\Omega \subseteq X$, and $L \in Y$. For a function $f : \Omega \to Y$, it is said that the limit of f as x approaches p is L (i.e., $f(x) \to L$ as $x \to p$) and write

$$\lim_{x \to p} f(x) = L$$

if the following property holds:

- For every open neighbourhood V of L, there exists an open neighbourhood U of p such that $f(U \cap \Omega - \{p\}) \subseteq V$.

This last part of the definition can also be phrased "there exists an open punctured neighbourhood U of p such that $f(U \cap \Omega) \subseteq V$".

Note that the domain of f does not need to contain p. If it does, then the value of f at p is irrelevant to the definition of the limit. In particular, if the domain of f is $X - \{p\}$ (or all of X), then the limit of f as $x \to p$ exists and is equal to L if, for all subsets Ω of X with limit point p, the limit of the restriction of f to Ω exists and is equal to L.

Sometimes this criterion is used to establish the *non-existence* of the two-sided limit of a function on R by showing that the one-sided limits either fail to exist or do not agree. Such a view is fundamental in the field of general topology, where limits and continuity at a point are defined in terms of special families of subsets, called filters, or generalized sequences known as nets.

Alternatively, the requirement that Y be a Hausdorff space can be relaxed to the assumption that Y be a general topological space, but then the limit of a function may not be unique. In particular, one can no longer talk about *the limit* of a function at a point, but rather *a limit* or *the set of limits* at a point.

A function is continuous in a limit point p of and in its domain if and only $f(p)$ is *the* (or, in the general case, *a*) limit of $f(x)$ as x tends to p.

Evaluating Limits at Infinity for Rational Functions

There are three basic rules for evaluating limits at infinity for a rational function $f(x) = p(x)/q(x)$: (where p and q are polynomials):

- If the degree of p is greater than the degree of q, then the limit is positive or negative infinity depending on the signs of the leading coefficients;

- If the degree of p and q are equal, the limit is the leading coefficient of p divided by the leading coefficient of q;
- If the degree of p is less than the degree of q, the limit is 0.

If the limit at infinity exists, it represents a horizontal asymptote at $y = L$. Polynomials do not have horizontal asymptotes; they may occur with rational functions.

Limit of a Function of More Than one Variable

By noting that $|x-p|$ represents a distance, the definition of a limit can be extended to functions of more than one variable. In the case of a function $f: \mathrm{R}^2 \to \mathrm{R}$,

$$\lim_{(x,y)\to(p,q)} f(x,y) = L$$

if for every $\varepsilon > 0$ there exists a $\delta > 0$ such that for all (x,y) with $0 < ||(x,y)-(p,q)|| < \delta$, then $|f(x,y)-L| < \varepsilon$ where $||(x,y)-(p,q)||$ represents the Euclidean distance. This can be extended to any number of variables.

Sequential Limits

Let $f: X \to Y$ be a mapping from a topological space X into a Hausdorff space Y, $p \in X$ and $L \in Y$.

The sequential limit of f as $x \to p$ is L if, for every sequence (x_n) in X-$\{p\}$ which converges to p, the sequence $f(x_n)$ converges to L.

If L is the limit (in the sense above) of f as x approaches p, then it is a sequential limit as well, however the converse need not hold in general.

If in addition Y is metrizable, then L is the sequential limit of f as x approaches p if and only if it is the limit (in the sense above) of f as x approaches p.

Other Characterizations

Limit of a Function in Terms of Sequences: For functions on the real line, one way to define the limit of a function is in terms of the limit of sequences. In this setting:

if and only if for all sequences (with not equal to a for all n) converging to the sequence converges to. It was shown by

Sierpiñski in 1916 that proving the equivalence of this definition and the definition above, requires and is equivalent to a weak form of the axiom of choice. Note that defining what it means for a sequence to converge to requires the epsilon, delta method.

Limit of a Function in Non-standard Calculus

In non-standard calculus the limit of a function is defined by:

$$\lim_{x \to a} f(x) = L$$

if and only if for all $x \in \mathbb{R}^*$, $f^*(x) - L$ is infinitesimal whenever $x - a$ is infinitesimal. Here $\mathbb{R}^*$ are the hyperreal numbers and f^* is the natural extension of *f* to the non-standard real numbers. Keisler proved that such a hyperreal definition of limit reduces the quantifier complexity by two quantifiers.

On the other hand, Hrbacek writes that for the definitions to be valid for all hyperreal numbers they must implicitly be grounded in the ε-δ method, and claims that, from the pedagogical point of view, the hope that non-standard calculus could be done without ε-δ methods can not be realized in full. B@aszczyk et al. detail the usefulness of microcontinuity in developing a transparent definition of uniform continuity, and characterize Hrbacek's criticism as a "dubious lament".

Limit of a Function in Terms of Nearness

At the 1908 international congress of mathematics F. Riesz introduced an alternate way defining limits and continuity in concept called "nearness". A point *x* is defined to be near a set $A \subseteq \mathbb{R}$ if for every $r > 0$ there is a point $a \in A$ so that $|x - a| < r$. In this setting the

$$\lim_{x \to a} f(x) = L$$

If and only if for all $A \subseteq \mathbb{R}$, is near $f(A)$ whenever a is near A. Here $f(A)$ is the set $\{y \in \mathbb{R} \mid y = f(x)\}$. This definition can also be extended to metric and topological spaces.

Relationship to Continuity

The notion of the limit of a function is very closely related to the concept of continuity. A function f is said to be continuous at c if it is both defined at c and its value at c equals the limit of f as x approaches c:

$$\lim_{x \to c} f(x) = f(c).$$

If the condition $0 < |x - c|$ is left out of the definition of limit, then the resulting definition would be equivalent to requiring f to be continuous at c.

Properties

If a function f is real-valued, then the limit of f at p is L if and only if both the right-handed limit and left-handed limit of f at p exist and are equal to L.

The function f is continuous at p if and only if the limit of $f(x)$ as x approaches p exists and is equal to $f(p)$. If $f: M \to N$ is a function between metric spaces M and N, then it is equivalent that f transforms every sequence in M which converges towards p into a sequence in N which converges towards $f(p)$.

If N is a normed vector space, then the limit operation is linear in the following sense: if the limit of $f(x)$ as x approaches p is L and the limit of $g(x)$ as x approaches p is P, then the limit of $f(x) + g(x)$ as x approaches p is $L + P$. If a is a scalar from the base field, then the limit of $af(x)$ as x approaches p is aL.

If f is a real-valued (or complex-valued) function, then taking the limit is compatible with the algebraic operations, *provided* the limits on the *right* sides of the equations below exist (the last identity only holds if the denominator is non-zero). This fact is often called the algebraic limit theorem.

$$\begin{aligned}
\lim_{x \to p} \; (f(x) + g(x)) &= \lim_{x \to p} f(x) + \lim_{x \to p} g(x) \\
\lim_{x \to p} \; (f(x) - g(x)) &= \lim_{x \to p} f(x) - \lim_{x \to p} g(x) \\
\lim_{x \to p} \; (f(x) \cdot g(x)) &= \lim_{x \to p} f(x) \cdot \lim_{x \to p} g(x) \\
\lim_{x \to p} \; (f(x) / g(x)) &= \lim_{x \to p} f(x) / \lim_{x \to p} g(x)
\end{aligned}$$

In each case above, when the limits on the right do not exist, or, in the last case, when the limits in both the numerator and the denominator are zero, nonetheless the limit on the left, called an *indeterminate form*, may still exist—this depends on the functions f and g. These rules are also valid for one-sided limits, for the case p = ±H, and also for infinite limits using the rules

- $q + \infty = \infty$ for q ¹ -∞
- $q \times \infty = \infty$ if $q > 0$
- $q \times \infty = -\infty$ if $q < 0$
- $q / \infty = 0$ if $q \neq \pm\infty$.

Note that there is *no* general rule for the case q / 0; it all depends on the way 0 is approached. Indeterminate forms—for instance, 0/0, 0×H, ∞-∞, and H/H—are also not covered by these rules, but the corresponding limits can often be determined with L'Hôpital's rule or the Squeeze theorem.

Chain Rule

In general, the statement

$\lim_{y \to d} f(y) = e$ and

$\lim_{x \to c} g(x) = d \Rightarrow \lim_{x \to c} f(g(x)) = e$,

is not true. However, this "chain rule" does hold if, in addition, either $f(d) = e$ (i. e. f is continuous at d) or g does not take the value d near c (i. e. there exists a $\delta > 0$ such that if $0 < |x - c| < \delta$ then $|g(x) - d| > 0$).

Limits of special interest

$$\lim_{x \to 0} \frac{\sin x}{x} = 1$$

$$\lim_{x \to 0} \frac{1 - \cos x}{x} = 0$$

$$\lim_{x \to \infty} x \sin(\frac{c}{x}) = c$$

The first limit can be proven with the squeeze theorem. For $0 < x < \pi/2$:

$$\sin x < x < \tan x.$$

Dividing everything by sin(x) yields

$$1 < \frac{x}{\sin x} < \frac{\tan x}{\sin x}$$

$$1 < \frac{x}{\sin x} < \frac{1}{\cos x}$$

$$\lim_{x\to 0}\frac{1}{\cos x} = \frac{1}{1} = 1$$

$$\lim_{x\to 0}\frac{x}{\sin x} = 1$$

$$\lim_{x\to 0}\frac{\sin x}{x} = 1$$

L'Hôpital's Rule

This rule uses derivatives and has a conditional usage. (It can only be directly used on limits that "equal" 0/0 or ±∞/±∞. Other indeterminate forms require some algebraic manipulation usually involving setting the limit equal to y, taking the natural logarithm of both sides, and then applying l'Hôpital's rule.)

$$\lim_{x\to c}\frac{f(x)}{g(x)} = \lim_{x\to c}\frac{f'(x)}{g'(x)}$$

For example: $\lim_{x\to 0}\frac{\sin(2x)}{\sin(3x)} = \lim_{x\to 0}\frac{2\cos(2x)}{3\cos(3x)} = \frac{2\cdot 1}{3\cdot 1} = \frac{2}{3}.$

Summations and Integrals

A short way to write the limit $\lim_{n\to\infty}\sum_{i=s}^{n} f(i)$ is $\sum_{i=s}^{\infty} f(i)$.

A short way to write the limit $\lim_{x\to\infty}\int_a^x f(t)\,dt$ is $\int_a^\infty f(t)\,dt$.

A short way to write the limit $\lim_{x\to-\infty}\int_x^b f(t)\,dt$ is $\int_{-\infty}^b f(t)\,dt$

Infinitesimal

Infinitesimals have been used to express the idea of objects so small that there is no way to see them or to measure them. The insight with exploiting infinitesimals was that objects could still retain certain defined properties, such as angle or slope, even though these objects were quantitatively small.

The word *infinitesimal* comes from a 17th century Modern Latin coinage *infinitesimus*, which originally referred to the "infinite-th" item in a series. It was originally introduced around 1670 by either Nicolaus Mercator or Gottfried Wilhelm Leibniz.

In common speech, an infinitesimal object is an object which is smaller than any feasible measurement, but not zero in size; or, so small that it cannot be distinguished from zero by any available means. Hence, when used as an adjective, "infinitesimal" in the vernacular means "extremely small".

An infinitesimal object by itself is often useless and not very well defined; in order to give it a meaning it usually has to be compared to another infinitesimal object in the same context (as in a derivative) or added together with an extremely large (an infinite) amount of other infinitesimal objects (as in an integral).

Archimedes used what eventually came to be known as the Method of indivisibles in his work *The Method of Mechanical Theorems* to find areas of regions and volumes of solids. In his formal published treatises, Archimedes solved the same problem using the Method of Exhaustion. The 15th century saw the work of Nicholas of Cusa, further developed in the 17th century by Johannes Kepler, in particular calculation of area of a circle by representing the latter as an infinite-sided polygon. Simon Stevin's work on decimal representation of all numbers in the 16th century prepared the ground for the real continuum. Bonaventura Cavalieri's method of indivisibles led to an extension of the results of the classical authors.

The method of indivisibles related to geometrical figures as being composed of entities of codimension 1. John Wallis's infinitesimals differed from indivisibles in that he would decompose geometrical figures into infinitely thin building blocks of the same dimension as the figure, preparing the ground for general methods of the integral calculus. He exploited an infinitesimal denoted $1/\infty$ in area calculations.

The use of infinitesimals by Leibniz relied upon heuristic principles, such as the Law of Continuity: what succeeds for the finite numbers succeeds also for the infinite numbers and vice versa; and the Transcendental Law of Homogeneity that specifies procedures for replacing expressions involving inassignable quantities, by expressions involving only assignable ones. The 18th century saw routine use of infinitesimals by mathematicians such as Leonhard Euler and Joseph Lagrange. Augustin-Louis Cauchy exploited infinitesimals in defining continuity and an early form of a Dirac delta function. As Cantor and Dedekind were developing more abstract versions of Stevin's continuum, Paul du Bois-Reymond wrote a series of papers on infinitesimal-enriched continua based on growth rates of functions. Du Bois-Reymond's work inspired both Émile Borel and Thoralf Skolem. Borel explicitly linked du Bois-Reymond's work to Cauchy's work on rates of growth of infinitesimals. Skolem developed the first non-standard models of arithmetic in 1934. A mathematical implementation of both the law of continuity and infinitesimals was achieved by Abraham Robinson in 1961, who developed non-standard analysis based on earlier work by Edwin Hewitt in 1948 and Jerzy £oœ in 1955. The hyperreals implement an infinitesimal-enriched continuum and the transfer principle implements Leibniz's law of continuity. The standard part function implements Fermat's adequality.

History of the Infinitesimal

The notion of infinitesimally small quantities was discussed by the Eleatic School. The Greek mathematician Archimedes (c.287 BC–c.212 BC), in *The Method of Mechanical Theorems*, was the first to propose a logically rigorous definition of infinitesimals. His Archimedean property defines a number x as infinite if it satisfies the conditions $|x|>1$, $|x|>1+1$,

$|x|>1+1+1$, ..., and infinitesimal if $x \neq 0$ and a similar set of conditions holds for $1/x$ and the reciprocals of the positive integers. A number system is said to be Archimedean if it contains no infinite or infinitesimal members.

The Indian mathematician Bhâskara II (1114–1185) described a geometric technique for expressing the change in $\sin \theta$ in terms of $\cos \theta$ times a change in θ. Prior to the invention of calculus mathematicians were able to calculate tangent lines by the method Pierre de Fermat's method of adequality and René Descartes method of normals. There is debate among scholars as to whether the method was infinitesimal or algebraic in nature. When Newton and Leibniz invented the calculus, they made use of infinitesimals. The use of infinitesimals was attacked as incorrect by Bishop Berkeley in his work *The Analyst*. Mathematicians, scientists, and engineers continued to use infinitesimals to produce correct results. In the second half of the nineteenth century, the calculus was reformulated by Augustin-Louis Cauchy, Bernard Bolzano, Karl Weierstrass, Cantor, Dedekind, and others using the (ε, δ)-definition of limit and set theory. While infinitesimals eventually disappeared from the calculus, their mathematical study continued through the work of Levi-Civita and others, throughout the late nineteenth and the twentieth centuries, as documented by Philip Ehrlich (2006). In the 20th century, it was found that infinitesimals could serve as a basis for calculus and analysis.

First-order Properties

In extending the real numbers to include infinite and infinitesimal quantities, one typically wishes to be as conservative as possible by not changing any of their elementary properties. This guarantees that as many familiar results as possible will still be available. Typically *elementary* means that there is no quantification over sets, but only over elements. This limitation allows statements of the form "for any number x..." For example, the axiom that states "for any number x, $x + 0 = x$" would still apply. The same is true for quantification over several numbers, e.g., "for any numbers x and y, $xy = yx$." However, statements of the form "for any *set* S of numbers ..." may not carry over. This limitation on quantification is referred to as first-order logic.

It superficially seems clear that the resulting extended number system cannot agree with the reals on all properties that can be expressed by quantification over sets, because the goal is to construct a nonarchimedean system, and the Archimedean principle can be expressed by quantification over sets, but this is just plain wrong.

It is trivial to conservatively extend any theory including reals, including set theory, to include infinitesimals, just by adding a countably infinite list of axioms that assert that a number is smaller than 1/2, 1/3, 1/4 and so on. Similarly, the completeness property cannot be expected to carry over, because the reals are the unique complete ordered field up to isomorphism. This is also wrong, at least as a formal statement, since it presuming some underlying model of set theory.

We can distinguish three levels at which a nonarchimedean number system could have first-order properties compatible with those of the reals:

An *ordered field* obeys all the usual *axioms* of the real number system that can be stated in first-order logic. For example, the commutativity axiom $x + y = y + x$ holds.

A *real closed field* has *all* the first-order properties of the real number system, regardless of whether they are usually taken as axiomatic, for statements involving the basic ordered-field relations +, *, and ≤. This is a stronger condition than obeying the ordered-field axioms. More specifically, one includes additional first-order properties, such as the existence of a root for every odd-degree polynomial. For example, every number must have a cube root.

The system could have all the first-order properties of the real number system for statements involving *any* relations (regardless of whether those relations can be expressed using +, *, and ≤). For example, there would have to be a sine function that iswell defined for infinite inputs; the same is true for every real function.

Systems in category 1, at the weak end of the spectrum, are relatively easy to construct, but do not allow a full treatment of classical analysis using infinitesimals in the spirit of Newton and Leibniz. For example, the transcendental functions are defined in terms of infinite limiting processes, and therefore there is typically no way to define them in first-order logic. Increasing the analytic strength of the system by passing to categories 2 and 3, we find that the flavour of the treatment tends to become less constructive, and it becomes more difficult to say anything concrete about the hierarchical structure of infinities and infinitesimals.

Number Systems That Include Infinitesimals

Formal Series

Laurent series: An example from category 1 above is the field of Laurent series with a finite number of negative-power terms. For example, the Laurent series consisting only of the constant term 1 is identified with the real number 1, and the series with only the linear term x is thought of as the simplest infinitesimal, from which the other infinitesimals are constructed. Dictionary ordering is used, which is equivalent to considering higher powers of x as negligible compared to lower powers. David O. Tall refers to this system as the super-reals, not to be confused with the superreal number system of Dales and Woodin. Since a Taylor series evaluated with a Laurent series as its argument is still a Laurent series, the system can be used to do calculus on transcendental functions if they are analytic. These infinitesimals have different first-order properties than the reals because, for example, the basic infinitesimal x does not have a square root.

The Levi-Civita Field

The Levi-Civita field is similar to the Laurent series, but is algebraically closed. For example, the basic infinitesimal x has a square root. This field is rich enough to allow a significant amount of analysis to be done, but its elements can still be represented on a computer in the same sense that real numbers can be represented in floating point.

Transseries

The field of transseries is larger than the Levi-Civita field. An example of a transseries is:

$$e^{\sqrt{\ln \ln x}} + \ln \ln x + \sum_{j=0}^{\infty} e^{x} x^{-j},$$

where for purposes of ordering x is considered to be infinite.

Surreal Numbers

Conway's surreal numbers fall into category 2. They are a system that was designed to be as rich as possible in different sizes of numbers, but not necessarily for convenience in doing analysis. Certain transcendental functions can be carried over to the surreals, including logarithms and exponentials, but most, e.g., the sine function, cannot. The existence of any particular surreal number, even one that has a direct counterpart in the reals, is not known a priori, and must be proved.

The Hyperreal System

The most widespread technique for handling infinitesimals is the hyperreals, developed by Abraham Robinson in the 1960s. They fall into category 3 above, having been designed that way in order to allow all of classical analysis to be carried over from the reals. This property of being able to carry over all relations in a natural way is known as the transfer principle, proved by Jerzy £oœ in 1955. For example, the transcendental function sin has a natural counterpart *sin that takes a hyperreal input and gives a hyperreal output, and similarly the set of natural numbers N has a natural counterpart ${}^*\mathbb{N}$, which contains both finite and infinite integers. A proposition such as $\forall n \in \mathbb{N}, \sin n\pi = 0$ carries over to the hyperreals as $\forall n \in^* \mathbb{N}, {}^*\sin n\pi = 0$.

Superreals

The superreal number system of Dales and Woodin is a generalization of the hyperreals. It is different from the super-real system defined by David Tall.

Smooth Infinitesimal Analysis

Synthetic differential geometry or smooth infinitesimal analysis have roots in category theory. This approach departs from the classical logic used in conventional mathematics by denying the general applicability of the law of excluded middle — i.e., *not* ($a \neq b$) does not have to mean $a = b$. A *nilsquare* or *nilpotent* infinitesimal can then be defined. This is a number x where $x^2 = 0$ is true, but $x = 0$ need not be true at the same time. Since the background logic is intuitionistic logic, it is not immediately clear how to classify this system with regard to classes 1, 2, and 3. Intuitionistic analogues of these classes would have to be developed first.

Infinitesimal Delta Functions

Cauchy used an infinitesimal α to write down a unit impulse, infinitely tall and narrow Dirac-type delta function δ_α satisfying $\int F(x)\delta_\alpha(x) = F(0)$ in a number of articles in 1827. Cauchy defined an infinitesimal in 1821 (Cours d'Analyse) in terms of a sequence tending to zero. Namely, such a null sequence becomes an infinitesimal in Cauchy's and Lazare Carnot's terminology.

Modern set-theoretic approaches allow one to define infinitesimals via the ultrapower construction, where a null sequence becomes an infinitesimal in the sense of an equivalence class modulo a relation defined in terms of a suitable ultrafilter. The article by Yamashita (2007) contains a bibliography on modern Dirac delta functions in the context of an infinitesimal-enriched continuum provided by the hyperreals.

Logical Properties

The method of constructing infinitesimals of the kind used in nonstandard analysis depends on the model and which collection of axioms are used. We consider here systems where infinitesimals can be shown to exist.

In 1936 Maltsev proved the compactness theorem. This theorem is fundamental for the existence of infinitesimals as it proves that it is possible to formalise them. A consequence of this theorem is that if there is a number system in which it is true that for any positive integer n there is a positive number x such that $0 < x < 1/n$, then there exists an extension of that number system in which it is true that there exists a positive number x such that for any positive integer n we have $0 < x < 1/n$. The possibility to switch "for any" and "there exists" is crucial. The first statement is true in the real numbers as given in ZFC set theory : for any positive integer n it is possible to find a real number between $1/n$ and zero, but this real number will depend on n. Here, one chooses n first, then one finds the corresponding x.

In the second expression, the statement says that there is an x (at least one), chosen first, which is between 0 and $1/n$ for any n. In this case x is infinitesimal. This is not true in the real numbers (R) given by ZFC. Nonetheless, the theorem proves that there is a model (a number system) in which this will be true. The question is: what is this model? What are its properties? Is there only one such model?

There are in fact many ways to construct such a one-dimensional linearly ordered set of numbers, but fundamentally, there are two different approaches:

1) Extend the number system so that it contains more numbers than the real numbers.
2) Extend the axioms (or extend the language) so that the distinction between the infinitesimals and non-infinitesimals can be made in the real numbers themselves.

In 1960, Abraham Robinson provided an answer following the first approach. The extended set is called the hyperreals and contains numbers less in absolute value than any positive real number. The method may be considered relatively complex but it does prove that infinitesimals exist in the universe of ZFC set theory. The real numbers are called standard numbers and the new non-real hyperreals are called nonstandard.

In 1977 Edward Nelson provided an answer following the second approach.

The extended axioms are IST, which stands either for Internal Set Theory or for the initials of the three extra axioms: Idealization, Standardization, Transfer. In this system we consider that the language is extended in such a way that we can express facts about infinitesimals. The real numbers are either standard or nonstandard. An infinitesimal is a nonstandard real number which is less, in absolute value, than any positive standard real number.

In 2006 Karel Hrbacek developed an extension of Nelson's approach in which the real numbers are stratified in (infinitely) many levels i.e., in the coarsest level there are no infinitesimals nor unlimited numbers. Infinitesimals are in a finer level and there are also infinitesimals with respect to this new level and so on.

Infinitesimal Calculus

Infinitesimal calculus is the part of mathematics concerned with finding slope of curves, areas under curves, minima and maxima, and other geometric and analytic problems.

History

Founders: It was independently developed by Gottfried Leibniz and Isaac Newton starting in the 1660s. John Wallis exploited an infinitesimal he denoted $1/\infty$ in area calculations, preparing the ground for integral calculus. They drew on the work of such mathematicians as Isaac Barrow and René Descartes. Infinitesimal calculus consists of differential calculus and integral calculus, respectively used for the techniques of *differentiation* and *integration*.

Newton sought to remove the use of infinitesimals from his fluxional calculus, preferring to talk of velocities as in "For by the ultimate velocity is meant ... the ultimate ratio of evanescent quantities". Leibniz embraced the concept fully calling differentials "...an evanescent quantity which yet retains the character of that which is disappearing", and his notation for them is the current symbolism in calculus.

Further Development

In early calculus the use of infinitesimal quantities was unrigorous and was fiercely criticized by a number of authors, most notably Michel Rolle and Bishop Berkeley. Berkeley mocked infinitesimals in his book *The Analyst* in 1734. Kirsti Andersen (2011) showed that Berkeley's doctrine of the compensation of errors in *The Analyst* contains a logical circularity. Namely, Berkeley relies upon Apollonius's determination of the tangent of the parabola in Berkeley's determination of the derivative of the quadratic function. A recent study argues that the force of Berkeley's criticisms has been overestimated; that Leibniz's defence of infinitesimals is more firmly grounded than Berkeley's criticism thereof; and that Leibniz's system for differential calculus was free of logical contradictions.

Several mathematicians, including Maclaurin and d'Alembert, attempted to prove the soundness of using limits, but it would be 150 years later, through the work of Augustin

Louis Cauchy and Karl Weierstrass, where a means was finally found to avoid mere "notions" of infinitely small quantities, that the foundations of differential and integral calculus were made firm. Cauchy developed a versatile spectrum of foundational approaches, including a definition of continuity in terms of infinitesimals and a (somewhat imprecise) prototype of an (ε, δ)-definition of limit in the definition of differentiation. Weierstrass formalized the concept of limit and eliminated infinitesimals. Following the work of Weierstrass, it eventually became common to base calculus on limits instead of infinitesimal quantities. This approach formalized by Weierstrass came to be known as the *standard* calculus. Informally, the expression "infinitesimal calculus" became commonly used to refer to Weierstrass' approach but has become something of a dead metaphor.

Modern Infinitesimals

After many years of the infinitesimal approach to calculus having fallen into disuse other than as an introductory pedagogical tool, use of infinitesimal quantities was finally given a rigorous foundation by Abraham Robinson in the 1960s. Robinson's approach, called non-standard analysis, uses technical machinery from mathematical logic to create a theory of hyperreal numbers that interpret infinitesimals in a manner that allows a Leibniz-like development of the usual rules of calculus.

Standard Calculus

It is based on the approach that took Weierstrass, replacing infinitesimals by limits. Limits describe the value of a function at a certain input in terms of its values at nearby input. They capture small-scale behaviour, just like infinitesimals, but use the ordinary real number system.

In this treatment, calculus is a collection of techniques for manipulating certain limits. Infinitesimals get replaced by very small numbers, and the infinitely small behaviour of the function is found by taking the limiting behaviour for smaller and smaller numbers. Limits are the easiest way to provide rigorous foundations for calculus, and for this reason they are the standard approach.

Non-standard Calculus

Based on Robinson's approach to infinitesimals. Calculations with infinitesimals were widely replaced with the (ε, δ)-definition of limit starting in the 1870s. For almost one hundred years thereafter, mathematicians like Richard Courant viewed infinitesimals as being naive and vague or meaningless.

Contrary to such views, Abraham Robinson showed in 1960 that infinitesimals are precise, clear, and meaningful, building upon work by Edwin Hewitt and Jerzy £oœ. According to Jerome Keisler, "Robinson solved a three hundred year old problem by giving a precise treatment of infinitesimals. Robinson's achievement will probably rank as one of the major mathematical advances of the twentieth century."

Smooth Infinitesimal Analysis

This is a mathematically rigorous reformulation of the calculus in terms of infinitesimals. Based on the ideas of F. W. Lawvere and employing the methods of category theory, it views all functions as being continuous and incapable of being expressed in terms of discrete entities. As a theory, it is a subset of synthetic differential geometry.

Chapter 12

Continuous Functions

Discontinuous Function

The simplest discontinuity occurs at those points at which a function has a jump, that is, at which the function has a definite limit as x tends to the point from the right and a definite limit as x tends to the point from the left, while these two limits are different. It does not matter whether or how the function is defined at the point of discontinuity itself.

For example, the function $f(x)$ defined by the equations

$$f(x)=0 \text{ for } x^2>1, f(x)=1 \text{ for}$$

$$x^2>1, f(x)=\frac{1}{2} \text{ for } x^3=1$$

has discontinuities at the points $\xi = 1$ and $\xi = -1$. The limits on approaching these points from the right and from the left hand side differ by 1, and the values of the function at these points agree with neither limit, but are equal to the arithmetic mean of the two limits. It may be noted in passing that our function can be represented, using the idea of a limit, by the expression

$$f(x) = \lim_{x\to\infty} \frac{1}{1+x^{2n}}$$

For, if $x < 1$, that is, if x lies in the interval $-1 < < 1$, the numbers x^{2n} will have the limit 0, and the function will have the value 1. However, if $x > 1$, as n increases, x^{2n} will increase beyond all bounds; our function will then have the value 0.

These are discontinuities which are exhibited by the functions $1/x$ or l/x at the point $\xi = 0$; as x ξ, the absolute value $| f(x) |$ of the function increases beyond all bounds.

In the case of $1/x$, the function increases numerically beyond all bounds through positive and through negative values, respectively, as x approaches the origin from the right and from the left.

On the other hand, the function l/x has for $x = 0$ an infinite discontinuity at which its value becomes positive from both sides. The function $y = 1/(x - 1)$, shown in Figure, has infinite discontinuities both at $x = 1$ and at $x = -1$.

Finally, we shall illustrate by an example another type of discontinuity in which there does not exist a limit from the right or from the left. Consider the function

$$y = \sin\frac{1}{x}$$

defined for all non-zero values of x. This function takes all values between -1 and $+1$ as the number $1/x$ ranges through the values from $(2n - 1/2)\pi$ to $(2n+1/2)\pi$, no matter what value has n. At the points $x=2/(4n-1)\pi$, the function will have the value -1, at the points $x = 2/(4n+1)\pi$ the value $+1$.

On the contrary, it oscillates backwards and forwards an infinite number of times, the magnitude of these oscillations becoming as small as we please as the origin is approached. This example shows us that even the simple idea of continuity admits all sorts of remarkable possibilities foreign to our naive intuition. There is one important fact which must be taken into account, if we are to give our ideas greater precision.

It may happen that at a certain point a function is not defined by the original law, as for example at the point $x = 0$ in the last two examples discussed. We have then the right to extend the definition of the function by assigning to it any desired value at such a point. However, in the last example, we can extend the definition in such a way that the function also remains continuous at that point, namely, by setting $y = 0$ when $x = 0$. This can be done whenever both the limits from the left and from the right exist and are equal to each other; we need then only make the value of the function at the point in question equal to these limits, in order to make the function continuous there. In the case of the function $y=\sin 1/a$, this cannot be done.

One Sided Limits

Although sometimes we get results directly, it is usually helpful to have a larger range of techniques. We give one here and more in section.

Definition: Say that $\lim\limits_{x\to a-} f(x) = l$, or that f has a limit from the left iff given $\in > 0$, there is some $\delta > 0$ such that whenever $a - \delta < x < a$, then $| f(x) - f(a) | < \in$. There is a similar definition of "limit from the right", writen as $\lim\limits_{x\to a+} f(x) = l$

Example: Define $f(x)$ as follows:

$$f(x) = \begin{cases} 3-x & \text{if } x<2 \\ 2 & \text{if } x<2; \\ x/2 & \text{if } x>2 \end{cases}$$

Calculate the left and right hand limits of $f(x)$ at 2.

Solution: As $x \to 2-$, $f(x) = 3 - x \to 1+$, so the left hand limit is 1. As $x \to 2+$, $f(x) = x/2 \to 1+$, so the right hand limit is 1. Thus the left and right hand limits agree (and disagree with $f(2)$, so f is not continuous at 2). Note our convention: If $f(x) \to 1$ and always $f(x) \geq 1$ as $x \to 2-$, we say that $f(x)$ tends to 1 from above, and write $f(x) \to 1+$ etc.

Proposition: If $\lim\limits_{x\to a} f(x)$ exists, then both one sided limts exist and are equal. Conversely, if both one sided limits exits and are equal, then $\lim\limits_{x\to a} f(x)$ exists. This splits the problem of finding whether a limit exists into two parts; we can look on either side, and check first that we have a limit, and second, that we get the same answer. For example both 1-sided limits exist, but are not equal. There is now an obvious way of checking continuity.

Continuity Test

Proposition: The function f is continuous at a iff both one sided limits exits and are equal to $f(a)$.

Example: Let $f(x) = \begin{cases} x^2 \text{ for } x \leq 1, \\ x \text{ for } x \geq 1. \end{cases}$

Show that f is continuous at 1.

Solution: We use the above criterion. Note that $f(1) = 1$. Also

$$\lim_{x \to 1-} f(x) = \lim_{x \to 1-} x^2 = 1$$

while $$\lim_{x \to 1+} f(x) = \lim_{x \to 1+} x = 1 = f(1).$$

so f is continuous at 1.

Example: Let $f(x) = |x|$. Then f is continuous in $\mathbb{R}$.

Solution: Note that if $x < 0$ then $|x| = -x$ and so is continuous, while if $x > 0$, then $|x| = x$ and so also is continuous. It remains to examine the function at 0. From these identifications, we see that,

$$\lim_{x \to 0-} |x| = 0+,$$

while $\lim_{x \to 0+} |x| = 0+$. Since $0+ = 0- = 0 = |0|$, by the $|x|$ is continuous at 0.

Results giving Continuity

Just as for sequences, building continuity directly by calculating limits soon becomes hard work. Instead we give a result that enables us to build new continuous functions from old ones, just as we did for sequences. Note that if f and g are functions and k is a constant, then $k.f$, $f + g$, fg and (often) f/g are also functions.

Proposition: Let f and g be continuous at a, and let k be a constant. Then $k.f$, $f + g$ and fg are continuous at f. Also, if $g(a) \neq 0$, then f/g is continuous at a.

Proof: We show that $f + g$ is continuous at a. Since, by definition, we have $(f + g)(a) = f(a) + g(a)$, it is enough to show that

$$\lim_{x \to \alpha} \left(f(x) + f(x) \right) 4 = f(a) + g(a).$$

Pick $\in > 0$; then there is some δ_1 such that if $| x - a | < \delta_1$, then $| f(x) - f(a) | </2$. Similarly there is some δ_2 such that if $| x - a | < \delta_2$, then $| g(x) - g(a) | < \in /2$. Let $= \min(\delta_1, \delta_2)$, and pick x with $| x - a | < \delta$. Then $| f(x) + g(x) - (f(a) + g(a)) | \leq | f(x) - f(a) | + | g(x) - g(a) | < \in /2 + \in /2 = \in$.

This gives the result. The other results are similar, but rather harder; for proofs.

Note: Just as when dealing with sequences, we need to know that f/g is defined in some neighbourhood of a. This can be shown using a very similar proof to the corresponding result for sequences.

Proposition: Let f be continuous at a, and let g be continuous at $f(a)$. Then gof is continuous at a

Proof: Pick $\in > 0$. We must find $\delta > 0$ such that if $| x - a | < \delta$, then $g(f(x)) - g(f(a)) | < \in$. We find δ using the given properties of f and g. Since g is continuous at $f(a)$, there is some $\delta_1 > 0$ such that if $| y - f(a) | < \delta_1$, then $| g(y) - g(f(a)) | < \in$.

Now use the fact that f is continuous at a, so there is some $\delta > 0$ such that if $| x - a | < \delta$, then $| f(x) - f(a) | < \delta_1$. Combining these results gives the required inequality.

Infinite Limits

There are many more definitions and results about limits. First one that is close to the sequence definition:

Definition: Say that $\lim_{x \to \infty} f(x) = l$ iff given $\in > 0$, there is some K such that whenever $x > K$, then $| f(x) - l | < \in$.

Example: Evaluate $\lim_{x \to \infty} \frac{x^2 + 3}{3x^2 + 2x + 1}$.

Solution: The idea here should be quite familiar from our sequence work. We use the fact that $1/x \to 0$ as $x \to \infty$. Thus

$$\frac{x^2 + 3}{3x^2 + 2x + 1} = \frac{1 + 3/x^2}{3 + 2/x + 1/x^2} \to \frac{1}{3} \text{ as } x \to \infty.$$

Definition Say that $\lim_{x \to \infty} f(x) = \infty$ iff given $L > 0$, there is some K such that whenever $x > K$, then $f(x) > L$.

The reason for working on proofs from the definition is to be able to check what results of this type are trivially true without having to find it in a book. For example

Proposition: Let $g(x) = 1/f(x)$. Then $g(x) \to 0+$ as x iff $f(x)$ as x. Let $y = 1/x$. Then $y \to 0+$ as $x \to \infty$; conversely, $y \to \infty$ as $x \to 0+$

Proof: Pick $\in > 0$. We show there is some K such that if $x > K$, then $0 < y < \in$; indeed, simply take $K = 1/\in$. The converse is equally trivial.

Closed Interval Continuity

So far our results have followed because f is continuous at a particular point. In fact we get the best results from assuming rather more. Indeed the results in this section are precisely why we are interested in discussing continuity in the first place.

Although some of the results are "obvious", they only follow from the continuity property, and indeed we present counterexamples whenever that fails. So in order to be able to use the following helpful results, we must first be able to check our functions satisfy the hypothesis of continuity. That task is of course what we have just been concentrating on.

Definition: Say that f is continuous on $[a, b]$ iff f is continuous on (a, b), and if, in addition, $\in \lim_{x \to a+} f(x) = f(a)$, and $\in \lim_{x \to b-} f(x) = f(b)$. We sometimes refer to f as being continuous on a compact interval. Such an f has some very nice properties.

Theorems on Boundedness

Theorem Let f be continuous on the compact interval $[a, b]$. Then there are constants M and m such that $m < f(x) < M$ for all $x \in [a, b]$. In other words, we are guaranteed that the graph of f is bounded both above and below.

Proof: This again uses the completeness of $\mathbb{R}$, and again no proof is offered. Note also that the hypotheses are all needed.

Theorem: Let f be continuous on the compact interval $[a, b]$. Then there are points x_0 and x_1 such that $f(x_0) < f(x) < f(x_1)$ for all $x \in [a, b]$.

In other words, we are guaranteed that the graph of f is bounded both above and below, and that these global extrema are actually attained.

Proof: This uses the completeness of $\mathbb{R}$, and follows in part from the previous result.

Consider the function $g(x) = (M - f(x))^{-1}$. This is clearly continuous. If there is some point at which $f(y) = M$, there is nothing to prove; otherwise $g(x)$ is defined everywhere, and is continuous, and hence bounded. This contradicts the fact that M was a least upper bound for f.

We have seen that any polynomial function $P(x)$ satisfies:

$$\lim_{x \to a} P(x) = P(a)$$

for all real numbers a. This property is known as continuity.

Definition: Let $f(x)$ be a function defined on an interval around a. We say that $f(x)$ is continuous at a iff

$$\lim_{x \to a} f(x) = f(a)$$

Otherwise, we say that $f(x)$ is discontinuous at a.

Note that the continuity of $f(x)$ at a means two things:

1. $\lim_{x \to a} f(x)$ exists,
2. And this limit is $f(a)$.

So to be discontinuous at a, means:

1. $\lim_{x \to a} f(x)$ does not exist,
2. or if $\lim_{x \to a} f(x)$ exists, then this limit is not equal to $f(a)$.

Basic Properties of Limits

Theorem. If $f(x)$ and $g(x)$ are continuous at a. Then

1. $f(x) + g(x)$ is continuous at a;
2. $\alpha f(x)$ is continuous at a, where is an arbitrary number;
3. $f(x)g(x)$ is continuous at a;
4. $\dfrac{f(x)}{g(x)}$ is continuous at a, provided $g(a) \neq 0$;
5. If $f(x)$ is positive, i.e. $f(x) \geq 0$, then $\sqrt{f(x)}$ is continuous at a;
6. If $f(x)$ is continuous at a and $g(x)$ is continuous at $f(a)$, then their composition $g o f(x)$ is continuous at a.

Remark: Many functions are not defined on open intervals. In this case, we can talk about one-sided continuity. Indeed, $f(x)$ is said to be continuous from the left at a iff

$$\lim_{x\to a-} f(x) = f(a)$$

and $f(x)$ is said to be continuous from the right at a iff

$$\lim_{x\to a+} f(x) = f(a)$$

Example: The function $f(x) = \sqrt{x}$ is defined for $x \geq 0$. So we can not talk about left-continuity of $f(x)$ at 0. But since we conclude that $f(x)$ is right-continuous at 0. This concept is also important for step-functions.

Example: Consider the function

$$f(x) = \begin{cases} x^3 + 2 & if \quad x < 2 \\ 5 & if \quad x = 2 \\ x^2 + 6 & if \quad x > 2 \end{cases}$$

The details are left to the reader to see

$$\lim_{x\to 2-} f(x) = \lim_{x\to 2-} x^3 + 2 = 10,$$

and

$$\lim_{x\to 2+} f(x) = \lim_{x\to 2+} x^2 + 6 = 10.$$

So we have

$$\lim_{x\to 2} f(x) = 10.$$

Since $f(2) = 5$, then $f(x)$ is not continuous at 2.

Chapter 13

Limit Theorem

We first met a sequence as a particularly easy sort of function, defined on $\mathbb{N}$, rather than $\mathbb{R}$. We now move to the more general study of functions. However, our earlier work wasn't a diversion; we shall see that sequences prove a useful tool both to investigate functions, and to give an idea of appropriate methods. Our main reason for being interested in studying functions is as a model of some behaviour in the real world.

Typically a function describes the behaviour of an object over time, or space. In selecting a suitable class of functions to study, we need to balance generality, which has chaotic behaviour, with good behaviour which occurs rarely. If a function has lots of good properties, because there are strong restrictions on it, then it can often be quite hard to show that a given example of such a function has the required properties.

Conversely, if it is easy to show that the function belongs to a particular class, it may be because the properties of that class are so weak that belonging may have essentially no useful consequences. It turns out that there are a number of "good" classes of functions which are worth studying.

In this chapter and the next ones, we study functions which have steadily more and more restrictions on them. Which means the behaviour steadily improves; and at the same time, the number of examples steadily decreases. A perfectly general function has essentially nothing useful that can be said about it; so we start by studying continuous functions, the first class that gives us much theory. In order to discuss functions sensibly, we often insist that we can "get a good look" at the behaviour of the function at a given point, so typically we restrict the domain of the function to be well behaved.

Definition: A subset U of $\mathbb{R}$ is open if given $a \in U$, there is some $\delta > 0$ such that $(a - \delta, a + \delta) \subseteq U$.

In fact this is the same as saying that given $a \in U$, there is some open interval containing a which lies in U. In other words, a set is open if it contains a neighbourhood of each of its points. We saw in that an open interval is an open set. This definition has the effect that if a function is defined on an open set we can look at its behaviour near the point a of interest, from both sides.

Functions And Mapping

Definition: Let A and B be any two non-empty sets. Let A = $\{p, q, r\}$ and $B = \{a, b, c, d\}$. Suppose by some rule or other, we assign to each element of A a 'unique' element of B. Let p be associated to a, q be associated to b, r be associated to c etc. The set $\{(p, a), (q, b), (r, c)\}$ is called a function from set A to set B. If we denote this set by 'f' then we write f: $A \rightarrow B$ which is read as "f" is a function of A to B or 'f' is a mapping from A to B.

Definition: Let A and B be two non-empty sets. Suppose there exists a rule of correspondence, denoted by f which associates to each member of A, a unique member of B. Then 'f' is called a function or a mapping from set A to set B.

It is denoted by f: $A \rightarrow$ B or $A \xrightarrow{f} B$

Note that:

1. The set A is called "domain" of 'f'
2. The set B is called "codomain" of 'f'
3. The element $y \in B$ which the function 'f' associates to the element $x \in A$ is denoted by $f(x)$ i.e.

   $y = f(x)$ and y is called the 'f' image of x or value of 'f' at x. The element x is also called as pre-image of y. Every element of A has a unique image but each element of B need not have as the image of an element in A. There can be more than one element of A which has the same image in B.

4. The set of these images in B is called the range of 'f' i.e. the range of a function as the images of its domain. We denote the range of $f: A \rightarrow B$ by f (A). Thus $f(A) = \{f(x): x \in A\} \Rightarrow f(A) \subseteq B$.
5. If A and B are any two non-empty sets, then a function f from A to B is a subset f of $A \times B$ satisfying the following conditions:

   a $\forall\ x \in A, (x, y) \in f$ for some $y \in B$

   b. $(x, y) \in f$ and $(x, y_1) \in f \Rightarrow y = y1$

   The first condition ensures that there is a rule which assigns to each element $x \in A$ some element $y \in B$. Thus each element in A will have an image. The second condition guarantees that the image is unique.

   In other words, a function is a set of ordered pair (x, y), such that for each first element 'x' there corresponds a unique element y.
6. The domain variable is often referred as the independent variable and the range variable is referred as the dependent variable.

Formula or Equation:Consider the function 'f' exhibited by the adjoining arrow diagram.

Let A = { 1, 2, 3, 4, 5}

B = { 5, 7, 9, 11, 13, 15}.

Note that, if we take any element $x \in A$ then the element of the co-domain set B related to x is obtained by adding 3 to twice of x. Applying this rule we get,

$f(1) = 2(1) + 3 = 5, f(2) = 2(2) + 3 = 7,$

$f(3) = 2(3) + 3 = 9$ etc.

In general we can write $f(x) = 2x + 3, x \in A$

This is the formula which exhibits the function 'f'. If we denote the value of f at x by, $y \in B$, we get $y = 2x + 3\ x \in A$. This becomes an equation which exhibits the function f.

Remarks:

1. If a function is exhibited by a formula, then using this formula we can find the range of the function.

2. Sometimes, the domain is not known or is not given. In such a case, the domain is taken as that set of elements at which the values of the function can be found. For example

   a. f(x) = $x^2 - 3x + 4$. In this case, the set R of all reals is taken as the domain of 'f'.

   b. f(x) $= \sqrt{16 - x^2}$ In order to have a real valued function, $16 - x^2 \geq 0$ i.e. $-4 \leq x \leq 4$ which the desired domain of 'f'

   c. $f(x) = \dfrac{x+7}{x-7}$ then for $x = 4$, $x + 7$ is not defined.

   Hence in this case, the domain is the $x - 4$ set R of reals except 4 i.e. R – {4}

Example: If A is the area of the circle. Describe this function.

Solution: If we call this function f then r is the independent variable and A becomes the dependent variable and we have $A = f(r) = \pi r^2$ for r ≥ 0. The domain must be stated with the constraint r ≥ 0 as we can't have a circle with negative radius.

Example: A relation is given as:

$$\{ (1, 1), (2, \frac{1}{2}), (3, \frac{1}{3}), (4, ¼), (5, \frac{1}{5})\}$$

is it a function ? Justify. If it is a function state its domain and range. Exhibit the relation by a formula.

Solution: Let,

$$A = \{ 1, 2, 3, 4, 5\} \text{ and } B = \{1, \frac{1}{2}, \frac{1}{3}, \frac{1}{4}, \frac{1}{5}\}$$

Every element in A occurs once and only once in the ordered pairs and is related to one and only one element of the set B, exhibiting a relation. Hence it is a function from the set A to the set B. Let us denote it by f. Then domain of A = The set A and the range of f = The set B. Also, the second member in each ordered pair is the reciprocal of the first member.

Thus if we denote any element of the set A by x and that of B by y then The formula is $f(x) = 1/x$ i.e. $y = 1/x\ x \in A$.

Example: A function f is given as $f(x) = x^2 - 6x + 5$. State its domain.

Find $f(-2), f(2), f(1), f(a + h), \dfrac{f(a+h)-f(a)}{h}, h \neq 0$

Solution: The domain of 'f' is the set R of all reals

$$f(x) = x^2 - 6x + 5 \Rightarrow f(-2) = (-2)^2 - 6(-2) + 5 = 21$$

$$f(2) = (2)^2 - 6(2) + 5 = -3, f(1)$$

$$= (1)^2 - 6(1) + 5 = 0$$

$$f(a + h) = (a + h)^2 - 6(a + h) + 5$$

$$= a^2 + 2ah + h^2 - 6a - 6h + 5$$

and,

$$= \frac{f(a+h)-f(a)}{h}$$

$$= \frac{(a^2+2ah+h^2-6a-6h+5)-(a^2-6h+5)}{h}$$

$$= \frac{h(2a-6+h)}{h}$$

Example: Find the range of each of the following functions:

1. $f(x) = 3x + 4$ for $-5 \leq x \leq 8$
2. $g(x) = 2x^2 + 3$ for $-4 \leq x \leq 3$
3. $\phi(x) = x^2 - 6x + 7$ for all $x \in R$

Solution:

1. The domain of ' f ' $= \{ x \mid x \in R, -5 \leq x \leq 8\}$

   Now $-5 \leq x \leq 8 \backslash 3(-5) \leq 3(x) \leq 3(8)$

   $\therefore -15 \leq 3x \leq 24 \therefore -15 + 4 \leq 3x + 4 \leq 24 + 4$

   $\therefore -11 \leq f(x) \leq 28$

The range of ' f ' = { $f(x) \in R, -11 \leq f(x) \leq 28$}

2. $-4 \leq x \leq 3 \Rightarrow 0 \leq x^2 \leq 16$

$\therefore 2\ (0) \leq 2\ (x) \leq 2\ (16) \therefore\ 0 + 3 \leq 2\ x^2 + 3 \leq 30 + 3$

$\therefore 3 \leq 9(x) \leq 35$

3. $h(x) = x^2 - 6x + 7 = x^2 - 6x + 9 - 9 + 7 = (x-3)^2 - 2$

Since $x \in R$, $(x - 3) \in R \Rightarrow (x - 3)2 \geq 0 \Rightarrow (x - 5)$ $2 - 2 \geq 0 - 2$

$\therefore h\ (x) \geq -2$. This is the range of '$h$'.

Example: Find the largest possible set of real numbers which can be the domain of

$$f(x) = \sqrt{1 - \tfrac{1}{x}}$$

Solution: Since,

$$\sqrt{1 - \tfrac{1}{x}}$$

is defined only when,

$$\sqrt{1 - \tfrac{1}{x}} \geq 0$$

i.e. $f(x)$ is defined for,

$$1 - \tfrac{1}{x} \geq 0$$

i.e.
$$f(x) = \frac{x-1}{x} \geq 0$$

Setting numerator and denominator separately equal to zero and solving, we get $x = 0$ and $x = 1$ (cut points)

Graph of expression of $f(x) > 0$

Graph of expression of $f(x) \geq 0$

$\therefore$ The domain is the set $= \{x \mid x < 0 \text{ or } x \geq 1\} = (-\infty, 0) \cup [1, \infty)$

Example: If $f\ (x) = \dfrac{x+5}{3x-4}$ and $t = \dfrac{5+4x}{3x-1}$

then show that $f\ (t) = x$

Solution:

$$f(x) = \frac{x+5}{3x-4} \backslash f(t) = \frac{t+5}{3t-4} \text{ putting } t = \frac{5+4x}{3x-1}$$

$$\therefore f(x) = \frac{\dfrac{5+4x}{3x-1}+5}{3\left(\dfrac{5+4x}{3x-1}\right)}$$

$$= \frac{\dfrac{5+4x+15x-5}{3x-1}}{\dfrac{5+12x-12x+4}{3x-1}}$$

$$\therefore f(t) = \frac{19x}{19} = x \backslash f(t) = x$$

Example: If $f(x) = x^2 - 2x + 3$, $x \in R$. Find the values of x for which $f(x) = f(3x - 1)$

Solution:

$$\begin{aligned} f(x) &= x^2 - 2x + 3 \Rightarrow f(3x-1) \\ &= (3x-1)^2 - 2(3x-1) + 3 \\ &= 9x^2 - 6x + 1 - 6x + 2 + 3 \\ &= 9x^2 - 12x + 6 \\ \therefore f(x) &= f(3x-1) \text{ gives us} \\ x^2 - 2x + 3 &= 9x^2 - 12x + 6 \\ 8x^2 - 10x + 3 &= 0 \\ (4x-3)(2x-1) &= 0 \\ 4x - 3 &= 0 \text{ or } 2x - 1 = 0 \\ \therefore x &= 3/4 \text{ or } x = 1/2 \end{aligned}$$

$\therefore$ The required values of x are 3/4 and 1/2

Concept of Limit

Convergence

If an infinite sequence of numbers $a_1, a_2, a_3, \ldots, a_n$, is given and if there is a number l such that every interval, however small, marked off about the point l, contains all the points a_n, except for at most a finite number, we say that the number l is the limit of the sequence a_1, a_2, ... or that the sequence a_1, a_2, ... converges to l; in symbols, $\lim_{x \to \infty} a_n = l$. Here we expressly remark that this includes the trivial case in which all the numbers a_n are equal to one another, and hence also coincide with the limit.

Instead of the above, we may use the following equivalent statement: If any positive number ε be assigned – however small – one can find a whole number N=N(ε) such that from N onward (i.e., for $n >$ N(ε)), it is always true that $|a_n - l| < \varepsilon$.. Of course, it is as a rule true that the bound $N(\varepsilon)$ will have to be chosen larger and larger with smaller and smaller values of ε; in other words, $N(\varepsilon)$ will increase beyond all bounds as ε tends to 0.

It is important to remember that every convergent sequence is bounded, that is, there corresponds to every sequence a_1, a_2, a_3, ..., for which a limit l exists, a positive number M, independent of n, such that for all the terms a_n of the sequence the inequality $|a_n| < M$ is valid. This theorem readily follows from our definition. We choose ε equal to 1; then there is an index N such that $|a_n - l| < 1$ for $n > N$. Let A be the largest among the numbers

$$|a_1 - l|, |a_2 - l|, \ldots, |a_n - l|$$

We can then put $M = [l] + A + 1$. Since, by the definition of A, the inequality $|a_n - l| < A + 1$ certainly holds for n= 1, 2, ..., N, while for $n > N$

$$|a_1 - l| < 1 \leqq A + 1$$

A sequence which does not converge is said to be divergent. If, as n increases, the numbers a_n increase beyond all bounds, we say that the sequence diverges to +} and, as we have already done occasionally, we write $\lim_{n \to \infty} a_n = \infty$.

Similarly, we write $\lim_{n \to \infty} a_n = -\infty$, if, as n increases, the numbers $-a_n$ increase beyond all bounds in the positive direction. But divergence may manifest itself in other ways, as, for example, in the case of the sequence $a_1 = -1, a_2 = +1, a_3 = -1, a_4 = +1, \ldots,$ the terms of which oscillate between two different values.

As regards convergence, the behaviour of a sequence is unaltered, if we omit a. finite number of the terms a_n. In what follows, we shall frequently use this, speaking of the convergence or divergence of series in which the term a_n is undefined for a finite number of values of n. In all the examples given above, the limit of the sequence considered is a known number.

If the concept of limit were to yield nothing more than the recognition that certain known numbers can be approximated as closely as we like by certain sequences of other known numbers, we should have gained very little.

The fruitfulness of the concept of limit in analysis rests essentially on the fact that limits of sequences of known numbers provide a means of dealing with other numbers, which are not directly known or expressible.

The whole of higher analysis consists of a succession of examples of this fact. The representation of the irrational numbers as limits of rational numbers may be regarded as a first example. In this section, we shall become acquainted with further examples. However, before we take up this subject, we shall make a few general, preliminary remarks.

Intrinsic Convergence

How can we tell that a given sequence of numbers a_1, a_2, a_3, ..., a_n, converges to a limit, even when we do not know beforehand what that limit is? This important question is answered by Cauchy's convergence test. It is sometimes referred to as the general principle of convergence.

We say that a sequence of numbers a_1, a_2, a_3, ..., a_n, is convergent if there corresponds to every arbitrarily selected small positive number ε a number $N = N(\varepsilon)$, which usually depends on ε, such that $|a_n - a_m| < \varepsilon$, provided that n and m are both at least equal to $N(\varepsilon)$. Cauchy's convergence test can then be expressed as follows:

Every intrinsically convergent sequence of numbers possesses a limit. The importance of Cauchy's test lies in the fact that it allows to speak of the limit of a sequence after considering the sequence itself without any further information about the limit itself. The converse of Cauchy's test is very easy to prove.

For, if the sequence a_1, a_2, ... tends to the limit l, then, by the definition of convergence, we have $|l - a_n| < \frac{\epsilon}{2}$ and $|l - a_m| < \frac{\epsilon}{2}$, where ε is a positive quantity as small as we please, provided only that both m and n are large enough, whence

$$|a_n - a_m| = |(l - a_m) - (l - a_n)| \leq |l - a_m| + |l - a_n| < \epsilon$$

Since ε can be chosen as small as we please, this inequality expresses our statement. Cauchy's test itself becomes intuitively obvious, if we think of the numbers as they are represented on the number axis. It then states that a sequence certainly has a limit, if after a certain point N all the terms of the sequence are restricted to an interval which can be made arbitrarily small by choosing N large enough.

Monotonic Sequence

The question whether a given sequence converges to a limit is particularly easy to answer when the sequence is a so-called monotonic sequence, that is, if either every number of the sequence is larger than the preceding number (monotonic increasing) or else every number is smaller than the preceding number (monotonic decreasing). We have the theorem:

Every monotonic increasing sequence the terms of which are bounded above (that is, lie below a fixed number) possesses a limit; similarly, every monotonic decreasing sequence the terms of which never drop below a certain fixed bound possesses a limit. Naturally, a convergent, monotonic, increasing sequence must tend to a limit which is greater than any term of the sequence, while for a convergent monotonic decreasing sequence the numbers tend to a limit which is smaller than any number of the sequence. Thus, for example, the numbers $1/n$ form a monotonic decreasing sequence with the limit 0, while the numbers $1 - 1/n$ form a monotonic increasing sequence with the limit 1. In many cases, it is convenient to replace the condition that a sequence shall increase monotonically by the weaker condition that its terms shall never decrease, in other words, to allow its successive terms to be equal to one another. We then speak of a monotonic non-decreasing sequence, or of a monotonic increasing sequence in the wider sense. Our theorem on limits remains true for such sequences as well as for sequences which are monotonic non-increasing or monotonic decreasing in the wider sense.

Limit with Variable Continuous

Hitherto, we have considered limits of sequences, that is, of functions of an integral variable n. However, the notion of limit frequently occurs in connection with the concepts of a continuous variable x and a function.

We say that the value of the function $f(x)$ tends to a limit l as x tends to ξ, or in symbols

$$\lim_{x \to \xi} f(x) = l,$$

if all the values of the function $f(x)$, for which x lies near enough to ξ, differ arbitrarily little from l. Expressed more precisely, the condition is:

If an arbitrarily small positive quantity ε is assigned, we can mark off about ξ an interval $|x - \xi| < \delta$ so small that there applies for every point x in this interval different from ξ itself the inequality $|f(x) - l| < \varepsilon$.

We expressly exclude here the equality of x and ξ. This is done purely for reasons of expediency, so as to have the definition in a form more convenient for applications, e.g., in the case where the function $f(x)$ is not defined at the point ξ, although it is defined for all other points in a neighbourhood of ξ.

If our function is defined or considered in a given interval only, for example, $\sqrt{1-x^2}$ in $-1 \leqq x \leqq 1$, we shall restrict the values of x to this interval. Thus, if ξ denotes an end-point of an interval, x is made to approach ξ by values on one side of ξ only (limit from the interior of the interval or one-sided limit).

As an immediate consequence of this definition, we have the fact:

If $\lim_{x \to \xi} f(x) = l$ and $x_1, x_2, \ldots, x_n$, is a sequence of numbers all different from ξ, but approaching ξ as a limit, then $\lim_{x \to \infty} f(x_n) = l$.

In fact, let ε be any positive number; we wish to show that for all values of n greater than a certain n_0 there applies the inequality.

By definition, there exists a $\delta > 0$ such that, whenever $|x - \xi| < \delta$, one has the inequality $|f(x_n) - l| < \in$.

Since $x_n < \xi$, the relation $|x - \xi| < \delta$ is satisfied for all sufficiently large values of n, and it follows for such values that $|f(x_n) - l| < \varepsilon$, as was to be proved. We shall now attempt to clarify this abstract definition by means of simple examples. Consider first the function

$$f(x) = \frac{\sin x}{x}$$

defined for x 0. We state that

$$\lim_{x \to 0} \frac{\sin x}{x} = 1$$

We cannot prove this statement simply by carrying out the passage to the limit in the numerator and denominator separately, because the numerator and denominator vanish when $x = 0$, and the symbol 0/0 has no meaning. We arrive at the proof as follows:

From Fig, we find by comparing the areas of the triangles OAB and OAC and the sector OAB that, if $0 < x < \pi/2$,

$$\sin x < x < \tan x \cdot$$

From this follows that, if $0 < |x| < \pi/2$,

$$1 < \frac{x}{\sin x} < \frac{1}{\cos x}$$

Hence the quotient $\sin x/x$ lies between the numbers 1 and cos x. We know that cos $x < 1$ as $x < 0$, whence the quotient sin x/ x can only differ arbitrarily little from 1, provided that x is near enough to 0. This is exactly what is meant by the equation which was to be proved. By the result just proved

$$\lim_{x\to 0}\frac{\tan x}{x} = \lim_{x\to 0}\frac{\sin x}{x}\lim_{x\to 0}\frac{1}{\cos x} = 1,$$

and also $$\lim_{x\to 0}\frac{1-\cos x}{x} = 0$$

This last result follows from the formula, valid for $0 < |x| < \pi/2$,

$$\frac{1-\cos x}{x} = \frac{(1-\cos x)(1+\cos x)}{x(1+\cos x)} = \frac{1-\cos^2 x}{x(1+\cos x)}$$

$$= \frac{\sin x}{x}.\frac{1}{1+\cos x}.\sin x.$$

As $x < 0$, the first factor on the right hand side tends to 1, the second one to 1/2 and the third one to 0, whence the product tends to 0, as has been stated.

$$\frac{1-\cos x}{x^2} = \left(\frac{\sin x}{x}\right)^2\frac{1}{1+\cos x}$$

$$\lim_{x\to 0}\frac{1-\cos x}{x^2} = \frac{1}{2}$$

Finally, consider the function $\sqrt{x^2}$, defined for all values of x. This function is never negative; it is equal to x for $x > 0$ and to $-x$ for $x < 0$. In other words $\sqrt{x^2} = |x|$.

Hence the function $\sqrt{x^2}/x$, defined for all non-zero values of x, has the value +1 when $x > 0$ and —1 when $x < 0$, whence the limit cannot exist, since arbitrarily near to 0 we can find values of x for which the quotient is +1 and other values for which it is –1.

In concluding this discussion on limits in connection with a continuous variable, we remark that it is, of course, possible to consider limiting processes in which the continuous variable x increases beyond all bounds. We state the example

$$\lim_{x\to\infty}\frac{x^2+1}{x^2-1} = \lim_{x\to\infty}\frac{1+\frac{1}{x^2}}{1-\frac{1}{x^2}} = 1$$

without further discussion. It signifies that the function on the left hand side differs arbitrarily little from 1, provided only that x is sufficiently large. In these examples, we have proceeded as if operations with limits obeyed the same laws in the case of continuous variables as in the case of sequences. That this is actually true, the reader himself can verify; the proofs are essentially the same as for limits of sequences.

One of the mysteries of Mathematics seems to be the concept of "infinity", usually denoted by the symbol. So what is ? It is simply a symbol that represents large numbers. Indeed, numbers are of three kinds: large, normal size, and small. The normal size numbers are the ones that we have a clear feeling for. For example, what does a trillion mean? That is a very large number. Also numbers involved in macro-physics are very large numbers. Small numbers are usually used in micro-physics. Numbers like 10^{-75} are very small. Being positive or negative has special meaning depending on the problem at hand.

The common mistake is to say that $-\infty$ is smaller than 0. While this may be true according to the natural order on the real line in term of sizes, $-\infty$ is big, very big! So when do we have to deal with and $-\infty$? Easy: whenever you take the inverse of small numbers, you generate large numbers and vice-versa. Mathematically we can write this as:

$$\frac{1}{0} = \pm\infty \text{ and } \frac{1}{\infty} = 0$$

Note that the inverse of a small number is a large number. So size-wise there is no problem. But we have to be careful about the positive or negative sign. We have to make sure we know whether a small number is positive or negative. 0+ represents small positive numbers while 0- represents small negative numbers. (Similarly, we will use e.g. 3+ to denote numbers slightly bigger than 3, and 3- to denote numbers slightly smaller than 3.) In other words, being more precise we have

$$\frac{1}{0+} = \pm\infty \text{ and } \frac{1}{0-} = -\infty$$

Remark: Do not $\pm\infty$ treat as ordinary numbers. These symbols do not obey the usual rules of arithmetic, for instance $\infty + 1 = \infty, \infty - 1 = \infty,\ 2, \infty = \infty$, etc.

Example: Consider the function

$$f(x) = \frac{2x+1}{x-1}$$

We have

$$\frac{2x+1}{x-1} = \frac{2+\frac{1}{x}}{1-\frac{1}{x}}$$

which implies

$$\lim_{x\to\pm\infty} f(x) = \frac{2+0}{1-0} = 2\,.$$

Note that when x gets closer to $\pm\infty$ (x gets large), then the points on the graph get closer to the horizontal line $y = 2$. Such a line is called a horizontal asymptote. In particular, we have

$$\lim_{x\to-\infty} \frac{a}{x^r} = 0 \; and \; \lim_{x\to+\infty} \frac{a}{x^r} = 0$$

for any number a, and any positive number r, provided x^r is defined. We also have

$$\lim_{x \to \infty} x^r = \infty$$

For $-\infty$, we have to be careful about the definition of the power of negative numbers. In particular, we have

$$\lim_{x \to -\infty} x^n = (-1)^n \infty$$

for any natural number n.

Many things grow at different rates which can be described by functions. Yet each of these functions of growth has some factor that allows it to grow only so much. An often used example is population growth. Populations (of people, animals, insects, bacteria, etc.) can continue to grow only as long as they have enough room and food. If the amount of food and space to live are constant, then as the population increases, there will be less for each individual to consume and less room to live and grow.

When the population exceeds a size that can survive with the given food and space it grows too big it will begin to decrease due to death. The population will level out around a number that can be sustained with the available resources, and stay around that number. Every population has a level point at which it can survive; a number that can be sustained by the available space, a limit.

Definition: Say that $f(x)$ tends to l as $x \to a$ iff given $\in >$ 0, there is some $\delta > 0$ such that whenever $0 < |x - a| < \delta$, then $|f(x) - l| < \in$.

Note that we *exclude* the possibility that $x = a$ when we consider a limit; we are *only* interested in the behaviour of f near a, but not at a. In fact this is very similar to the definition we used for sequences. Our main interest in this definition is that we can now describe continuity accurately.

Definition: Say that f is continuous at a if $\lim_{x \to \alpha} f(x) = f(a)$. Equivalently, f is continuous at a iff given $\in > 0$, there is some $\delta > 0$ such that whenever $|x - a| < \delta$, then $|f(x) - f(a)| < \in$.

Note that in the "epsilon – delta" definition, we no longer need exclude the case when $x = a$. Note also there is a good geometrical meaning to the definition. Given an error, there is *some* neighbourhood of a such that if we stay in that neighbourhood, then f is trapped within of its value $f(a)$. We shall not insist on this definition in the same way that the definition of the convergence of a sequence was emphasised. However, all our work on limts and continuity of functions can be traced back to this definition, just as in our work on sequences, everything could be traced back to the definition of a convergent sequence. Rather than do this, we shall state without proof a number of results which enable continuous functions both to be recognised and manipulated.

So you are expected to know the definition, and a few simply $\in -\delta$ proofs, but you can apply (correctly – and always after checking that any needed conditions are satisfied) the standard results we are about to quote in order to do required manipulations.

Definition: Say that f: *U(open)* $\rightarrow \mathbb{R}$ is continuous if it is continuous at each point $a \in U$.

Note:This is important. The function $f(x) = 1/x$ is defined on $\{x{:}x \neq 0\}$, and is a continuous function. We cannot usefully define it on a larger domain, and so, by the definition, it is continuous. This is an example where the naive "can draw it without taking the pencil from the paper" definition of continuity is not helpful.

Example: Let $f(x) \dfrac{x^3-8}{x-2} =$ for $x \neq 2$. Show how to define f (2) in order to make f a continuous function at 2.

Solution: We have

$$\frac{x^3-8}{x-2} = \frac{(x-2)\left(x^2+2x+4\right)}{(x-2)} = (x^2 + 2x + 4)$$

Thus $f(x) \rightarrow (2^2 + 2.2 + 4) = 12$ as $x \rightarrow 2$. So defining $f(2) = 12$ makes f continuous at 2, (and hence for all values of x).

Chapter 14

Infinite Series

Introduction

The sum of an infinite series $a_0 + a_1 + a_2 + \dots$ is the limit of the sequence of partial sums

$$S_n = a_0 + a_1 + a_2 + \dots + a_n$$

as $n \to \infty$, if that limit exists. If the limit exists and is finite, the series is said to converge; if it is infinite or does not exist, the series is said to diverge.

The easiest way that an infinite series can converge is if all the a_n are zero for n sufficiently large. Such a series can be identified with a finite sum, so it is only infinite in a trivial sense. However, infinite series of nonzero terms can also converge, which resolves the mathematical side of several of Zeno's paradoxes. The simplest case of a nontrivial infinite series is perhaps

$$1+\frac{1}{2}+\frac{1}{4}+\frac{1}{8}+\frac{1}{16}+\dots$$

It is possible to "visualize" its convergence on the real number line: we can imagine a line of length 2, with successive segments marked off of lengths 1, ½, ¼, etc. There is always room to mark the next segment, because the amount of line remaining is always the same as the last segment marked: when we have marked off ½, we still have a piece of length ½ unmarked, so we can certainly mark the next ¼.

This argument does not prove that the sum is *equal* to 2 (although it is), but it does prove that it is *at most* 2. In other words, the series has an upper bound. This series is a geometric series and mathematicians usually write it as:

$$\sum_{n=0}^{\infty} 2^{-n} = 2.$$

An infinite series is formally written as

$$\sum_{n=0}^{\infty} a_n$$

where the elements a_n are real (or complex) numbers. We say that this series converges to S, or that its sum is S, if the limit

$$\lim_{N \to \infty} \sum_{n=0}^{N} a_n$$

exists and is equal to S. If there is no such number, then the series is said to *diverge*.

Formal Definition

Mathematicians usually study a series as a *pair* of sequences: the sequence of terms of the series: $a_0, a_1, a_2, \ldots$ and the sequence of partial sums $S_0, S_1, S_2, \ldots$, where $S_n = a_0 + a_1 + \ldots + a_n$. The notation

$$\sum_{n=0}^{\infty} a_n$$

represents the above sequence of partial sums, which is always well defined, but which may or may not converge. In the case of convergence (that is, when the sequence of partial sums S_N has a limit), the notation is also used to denote the *limit* of this sequence.

To make a distinction between these two completely different objects (sequence vs. numerical value), one may sometimes omit the limits (atop and below the sum's symbol) in the former case, although it is usually clear from the context which one is meant. Also, different notions of convergence of such a sequence do exist (absolute convergence, summability, etc). In case the elements of the sequence (and thus of the series) are not simple numbers, but, for example, functions, still more types of convergence can be considered. Mathematicians extend this idiom to other, equivalent notions of series. For instance, when we talk about a recurring decimal, we are talking, in fact, just about the series for which it stands (0.1 + 0.01 + 0.001 + ...). But because these series always converge to real numbers (because of what is called the completeness property of the real numbers), to talk about the series in this way is the same as to talk about the numbers for which they stand. In particular, it should offend no sensibilities if we make no distinction between 0.111... and $^1/_9$. Less clear is the argument that $9 \times 0.111... = 0.999... = 1$, but it is not untenable when we consider that we can formalize the proof knowing only that limit laws preserve the arithmetic operations.

The geometric series, Taylor series and several special examples, which we have already met, suggest that we may well study from a rather more general point of view those limiting processes under the heading of summation of infinite series. By its nature, any limiting value $S = \lim_{n \to \infty} s_n$ can be written as an infinite series, because, if n takes the values 1, 2, 3,..., we need only set $a_n = s_n - s_{n-1}$ for $n > 1$ and $a_1 = s_1$, in order to obtain $s_n = a_1 + a_2 + ... + a_n$, and the value S thus appears as the limit of s_n - the sum of n terms as n increases. We express this fact by saying that S is the sum of the infinite series $a_1 + a_2 + a_3 + ...$

Thus, an infinite series is simply a way of representing a limit, where each successive approximation is found from the preceding one by addition of one more term. In principle, the expression of a number as a decimal is merely the representation of a number a in the form of an infinite series $a = a_1 + a_2 + a_3 + ...$ where, if $0 < a < 1$, the term a_n is put equal to $a_n 10^{-n}$ and a_n is a whole number between 0 and 9, inclusively.

Since every limiting value can be written in the form of an infinite series, it may seem that a special study of series is superfluous. However, in many cases, it happens that limiting values occur naturally in the form of infinite series, which often exhibit particularly simple laws of formation. Of course, it is not true that every series has an easily recognizable law of formation. For example, the number p can certainly be represented as a decimal, yet we know no simple law which enables us to state the value of an arbitrary digit, say its 7000-th decimal. However, if we set aside the representation of π by a decimal and consider instead Gregory's series, we have an expression with a perfectly clear general law of formation.

Infinite series, in which approximations to the limit are found by repeated addition of new terms, are analogous to infinite products, in which the approximations to the limit arise from repeated multiplication by new factors. However, we shall not go deeply into the theory of infinite products.

The Concepts of Convergence and Divergence

The Fundamental Ideas: We consider an infinite series the general term of which we denote by a_n; the series then has the form $a_1 + a_2 + \ldots = \sum_{\nu=1}^{\infty} a_\nu$.

The symbol with the summation sign on the right hand side is merely an abbreviated way of writing the expression on the left hand side.

For formal reasons, we include the possibility that certain of the numbers a_n may be zero. If all a_n vanish onwards from a number N (i.e., when $n > N$), we speak of a terminating series.

If, as n increases, the n-th partial sum $s_n = a_1 + a_2 + \ldots + a_n$ $= \sum_{\nu=1}^{n} a_y$ approaches a limit $S = \lim_{n\to\infty} s_n$ we say that the series converges, and otherwise that it diverges. In the first case, we call S the sum of the series. We have already met with many examples of convergent series; for instance, the $1 + q + q^2 + \ldots$, geometric series which converges to the sum $1/(1-q)$ for $q < 1$, Gregory's series, the series for log 2, the series for e, and others converge. In the language of infinite series, Cauchy's convergence test is expressed as follows:

A necessary and sufficient condition for the convergence of a series is that the number $|s_m - s_n| = |a_{n+1} + a_{n+2} + \ldots + a_m|$ becomes arbitrarily small if m and n are chosen sufficiently large ($m > n$).

In other words: A series converges if and only if the following condition is met: Given a positive number ε, however small, it is possible to choose an index $N=N(\varepsilon)$, which, in general, increases beyond all bounds as $\varepsilon \succ 0$, in such a way that the $|s_m - s_n|$ is less than e, provided only that $m > N$ and $n > N$. We can make the meaning of the convergence test clearer by considering the geometric series with $q = 1/2$. If we choose $\varepsilon = 1/10$, we need only take $N = 4$, because,

$$\left|s_m - s_n\right| = \frac{1}{2^n} + \ldots + \frac{1}{2^{m-1}}$$

$$= \frac{1}{2^{n-1}}\left(\frac{1}{2} + \frac{1}{2^2} + \ldots + \frac{1}{2^{m-n}}\right) < \frac{1}{2^{n-1}}$$

and $\frac{1}{2^{n-1}} < \frac{1}{10}, if\ n > 4$.

If we choose ε equal to 1/100, it is sufficient to take 7 as the corresponding value of N, as is easily verified. Obviously, it is a necessary condition for the convergence of a series that $\lim_{n\to\infty} a_n = 0$. Otherwise, undoubtedly, the convergence criterion cannot be fulfilled. However, this necessary condition is by no means sufficient for convergence; on the contrary, it is easy to find infinite series the general term a_n of which approaches 0 as n increases, yet their sum does not exist, because the partial sum s_n increases without limit as n increases.

An example of this is the series $1 + \frac{1}{\sqrt{2}} + \frac{1}{\sqrt{3}} + \ldots + \frac{1}{\sqrt{n}} + \ldots$

with the general term $\frac{1}{\sqrt{n}}$. We immediately see that

$$s_n > \frac{1}{\sqrt{n}} + \ldots + \frac{1}{\sqrt{n}} = \frac{n}{\sqrt{n}} = \sqrt{n}$$

The n-th partial sum increases beyond all bounds as n increases, whence the the series diverges The same is true for the classical example of the harmonic series $1 + \frac{1}{2} + \frac{1}{3} + \frac{1}{4} + \ldots$, for which,

$$a_{n+1} + \ldots + a_{2n} = \frac{1}{n+1} + \ldots + \frac{1}{2n} > \frac{1}{2n} + \ldots + \frac{1}{2n} = \frac{1}{2}$$

Since n and $m = 2n$ can be taken as large as we please, the series diverges, because Cauchy's test is not fulfilled; in fact, the n-th partial sum tends obviously to infinity, since all the terms are positive. On the other hand, the series, formed from the same numbers with alternating signs,

$$1 - \frac{1}{2} + \frac{1}{3} - \frac{1}{4} + \frac{1}{5} - +.. + \frac{(-1)^{n-1}}{n} + \ldots$$

converges and has the sum log 2.

It is by no means true that in every divergent series s_n tends to +1 or –1. Thus, in the case of the series 1 – 1 + 1 – 1 + 1 + –..., we see that the partial sum s_n has alternately the values 1 and 0, and, on account of this backwards/ forwards oscillation neither approaches a definite limit nor increases numerically beyond all bounds. As the convergence and divergence of an infinite series is concerned, the following fact which, though self-evident, is very important should be noted:

The convergence or divergence of a series is not changed by insertion or removal of finite numbers of terms. As far as convergence or divergence is concerned, it does not matter in the least whether we begin the series at the term a_0, or a_1, or a_5, or at any other arbitrarily chosen term.

Absolute Convergence and Conditional Convergence:. The series $1 + \frac{1}{2} + \frac{1}{3} + \frac{1}{4} \ldots$ diverges; but if we change the sign of every second term, the resulting series converges. On the other hand, the geometric series 1-q+q converges and has the sum l/(l + q), provided that $0 < q < 1$; on converting all the signs to +, we obtain the series $1 + q + q^2 + q^3 + \ldots$, which also converges and has the sum l/(l - q).

There appears here a distinction which we must examine more closely. With a series all terms of which are positive, there are only two possible cases; either it converges or the partial sum increases beyond all bounds with n. In fact, the partial sums, being a monotonic increasing sequence, must converge, if they remain bounded. Convergence occurs, if the terms approach zero sufficiently rapidly as n increases; on the other hand, divergence occurs, if the terms do not approach zero at all or if they approach zero too slowly.

However, in series, where some terms are positive and others negative, the changes of sign may bring about convergence, since a too great increase in the partial sums, due to the positive terms, is compensated by negative terms, as a final result of which a definite limit is approached.

In order to grasp this fact better with a series $\sum_{\nu=1}^{\infty} a_\nu$, which has positive and negative terms, we compare it with the series which has the same terms all with positive signs, that is,

$$|a_1| + |a_2| + \ldots = \sum_{\nu=1}^{\infty} |a_\nu|$$

If this series converges, then for sufficiently large values of n and $m > n$, the expression $|a_1| + |a_{n+2}| + \ldots + |a_m|$ will certainly be as small as we please; on account of the relation $|a_{n+1} + \ldots + a_m| \leq |a_{n+1}| + \ldots + |a_m|$, the expression on the left hand side is also arbitrarily small, whence the original series $\sum_{\nu=1}^{\infty} a_\nu$ converges. In this case, the original series is said to be absolutely convergent. Its convergence is due to the numerical smallness of its terms and does not depend on the change of the signs.

On the other hand, if the series with all the terms taken positively diverges and the original series still converges, we say that the original series is conditionally convergent. Conditional convergence results from the terms of opposite signs compensating one another.

For conditional convergence, Leibnitz's convergence test is frequently of use: If the terms of a series have alternating sign and, moreover, their absolute values $|a_n|$ tend monotonically to 0 (so that $|a_{n+1}| < |a_n|$), the series $\sum_{\nu=1}^{\infty} a_\nu$ converges.

In the proof, we assume that $a_1 > 0$, which does not essentially limit the generality of the argument, and write our series in the form $b_1 - b_2 + b_3 - +\ldots$, where all the terms b_n are now positive, b_n tends to 0 and the condition $b_{n+1}<b_n$ is satisfied. If we bracket the terms together in the two ways $b_1 - (b_2 - b_3) - (b_4 + b_5) -\ldots$ and $(b_1 - b_2) + (b_3 + b_4) + (b_5 + b_6) +\ldots$

we see at once that the partial sums satisfy the two relations:

$$s_1 > s_2 > s_3 >\ldots > s_{2m+1} >\ldots$$

$$s_2 < s_4 < s_6 <\ldots < s_{2m} <\ldots$$

On the other hand, $s_{2n} < s_{2n+1} < s_1$ and $s_{2n+1} > s_{2n} > s_2$, whence the odd partial sums s_1, s_3,... form a monotonic decreasing sequence, which in no case drops below the value s_2 and this sequence has a limit L.

The even partial sums s_2, s_4,... likewise form a monotonic increasing sequence the terms of which in no case exceed the fixed number s_1, whence this sequence must have a limiting value L'. Since the numbers s_{2n} and s_{2n+1} differ only by the number b_{2n+1}, which approaches 0 as n increases, the limiting values L and L' are equal to each other. In other words, the even and the odd partial sums

In conclusion, we make another general remark about the fundamental difference between absolute convergence and conditional convergence. We consider a convergent series $\sum_{\nu=1}^{\infty} a_\nu$ and denote the positive terms of the series by $p_1, p_2, p_3,\ldots$ and the negative terms by $-q_1$, $-q_2$, $-q_3$,... If we form the n-th partial sum $s_n = \sum_{\nu=1}^{n} a_\nu$ of the given series, a certain nmnber, say, n' positive terms and a certain number, say n'' negative terms must appear, where $n' + n''=n$. Moreover, if the number of positive terms as well as that of negative terms in the series is infinite, then the two numbers n'' and n'' will increase beyond all bounds with n. We see immediately that the partial sum s_n

is simply equal to the partial sum $\sum_{\nu=1}^{n'} p_\nu$ of the positive terms of the series plus the partial sum $-\sum_{\nu=1}^{n''} q_\nu$ of the negative terms.

If the given series converges absolutely, then both the series of positive terms $\sum_{\nu=1}^{\infty} p_\nu$ and the series of the absolute values of the negative terms $\sum_{\nu=1}^{\infty} q_\nu$ certainly converge. In fact, as m increases, the partial sums $\sum_{\nu=1}^{m} p_\nu$ *and* $\sum_{\nu=1}^{m} q_\nu$ are monotonic non-decreasing with the upper bound $\sum_{\nu=1}^{\infty} |a_\nu|$.

The sum of an absolutely convergent series is then simply equal to the sum of the series consisting only of the positive terms plus the sum of the series consisting of the negative terms, or, in other words, it is equal to the difference of the two series with positive terms.

However, $\sum_{\nu=1}^{n} a_\nu = \sum_{\nu=1}^{n'} p_\nu - \sum_{\nu=1}^{n''} q_\nu$; as n increases, n' and n'' must also increase beyond all bounds, and the limit of the left hand side must therefore be equal to the difference of the two sums on the right hand side.

If the series contains only a finite number of terms of one particular sign, the facts are correspondingly simplified. On the other hand, if the series does not converge absolutely, but does converge conditionally, then both the series $\sum_{\nu=1}^{\infty} p_\nu$ and $\sum_{\nu=1}^{\infty} q_\nu$ must be divergent, because, if both were convergent, the series would converge absolutely, contrary to our hypothesis. If only one diverges, say $\sum_{\nu=1}^{\infty} p_\nu$, and the other converges, then separation into positive and negative parts, $s_n = \sum_{\nu=1}^{n'} p_\nu - \sum_{\nu=1}^{n''} q_\nu$, shows that the series could not converge;

in fact, as n increases, n' and $\sum_{\nu=1}^{n'} p_\nu$ would increase beyond all bounds, while the term $\sum_{\nu=1}^{n''} q_\nu$ would approach a definite limit, so that the partial sum s_n would increase beyond all bounds. Hence, we see that a conditionally convergent series cannot be thought of as the difference of two convergent series, one consisting of its positive terms and the other consisting of the absolute values of its negative terms. Closely connected with this fact, there arises another difference between absolutely and conditionally convergent series which we shall now mention briefly.

Rearrangement of Terms

It is a property of finite sums that we may change the order of the terms or, as we say, rearrange the terms at will without changing their values.

There arises the question: What is the exact meaning of a change of the order of terms in an infinite series and does such a rearrangement leave the value of the sum unchanged? While in the case of finite sums there arises no difficulty, for example, in adding the terms in reverse order, in the case of infinite series such a possibility does not exist; there is no last term with which to begin.

Now a change of the order of terms in an infinite series can only mean this: A series $a_1 + a_2 + a_3 + \ldots$ is rearranged into a series $b_1 + b_2 + b_3 + \ldots$, provided that every term a_n of the first series occurs exactly once in the second series and conversely. For example, the amount by which a_n is displaced may increase beyond all bounds as n does; the only point is that it must appear somewhere in the new series.

If some of the terms are moved to later positions in the series, other terms must, of course, be moved to earlier positions. For example, the series $1 + q + q^2 + q^4 + q^3 + q^8 + q^7$

$+ q^6 + q^5 + q^{16} + \ldots$ is a rearrangement of the geometric series $1 + q + q^2 + \ldots$

As regards the change of order, there is a fundamental distinction between absolutely convergent and conditionally convergent series. In absolutely convergent series, rearrangement of terms does not affect the convergence and the value of the sum of the series is not changed, exactly as in the case of finite sums. On the other hand, in conditionally convergent series, the value of the sum of the series can be changed at will by suitable rearrangement of the series, and the series can even be made to diverge, if so desired. The first of these facts, referring to absolutely convergent series, is easily established. Let us assume, to begin with, that our series has only positive terms and consider the n-th partial sum $s_n = \sum_{\nu=1}^{n} a_\nu$. All the terms of this partial sum occur in the m-th partial sum $t_m = \sum_{\nu=1}^{m} b_\nu$ of the rearranged series, provided only that m is chosen large enough. Hence $t_m \geq s_n$.

On the other hand, we can determine an index n' so large that the partial sum $s_{n'} = \sum_{\nu=1}^{n'} a_\nu$ of the first series contains all the terms b_1, b_2, b_m. It then follows that $t_m \leq s_{n'} \leq A$, where A is the sum of the first series. Thus, for all sufficiently large values of m, we have $s_n \leq t_m \leq A$; since s_n can be made to differ from A by an arbitrarily small amount, it follows that the rearranged series also converges, in fact, to the same limit A as the original series. If the absolutely convergent series has both positive and negative terms, we may regard it as the difference of two series each of which has only positive terms. Since in the rearrangement of the original series each of these two series merely undergoes rearrangement and therefore converges to the same value as before, the same is true of the original series when rearranged. In fact, by the case just considered, the new series is absolutely convergent and is therefore the difference of the two rearranged series of positive terms. The fact just proved may seem to the beginner to be a triviality. The fact that it really does require a proof and that in this proof absolute convergence is essential can be demonstrated by an example of the opposite behaviour of conditionally convergent series. Consider the familiar series

$$1-\frac{1}{2}+\frac{1}{3}-\frac{1}{4}+\frac{1}{5}-\frac{1}{6}+\frac{1}{7}-\frac{1}{8}+-\ldots = \log 2.$$

Write below it the result of multiplication by the factor $1/2\,\frac{1}{2}-\frac{1}{4}+\frac{1}{6}-\frac{1}{8}+-\ldots=\frac{1}{2}\log 2$ and add the two series, combining the terms in vertical columns.

We thus obtain $1-\frac{1}{3}-\frac{1}{2}+\frac{1}{5}+\frac{1}{7}-\frac{1}{4}+\frac{1}{9}+\frac{1}{11}++\ldots=\frac{3}{2}\log 2$.

This last series can obviously be obtained by rearranging the original series and yet the value of the sum of the series has been multiplied by the factor 3/2. It is easy to imagine the effect of the discovery of this apparent paradox on the mathematicians of the Eighteenth Century, who were accustomed to operate with infinite series without regard to their convergence.

We shall give the proof of the theorem stated above concerning the change in the sum of a conditionally convergent series which arises from a change of order of its terms, although we shall have no occasion to make use of the result. Let p_1, p_2,... be the positive terms and $-q_1$, $-q_2$,... the negative terms of the series. Since the absolute value $|a_n|$ tends to 0 as n increases, the numbers p_n and q_n must also tend to 0 with n.

Moreover, as we have already seen, the sum $\sum_1^\infty p_\nu$ must diverge and the same is true for $\sum_1^\infty q_\nu$. This abbreviated notation for $\sum_1^\infty p_\nu$ and analogous expressions for other series will be used often in what follows.

We can now find easily a rearrangement of the original series which has an arbitrary number a as its limit. In order to be specific, let a be positive.

We then add the first n_1 positive terms, just enough to ensure that the sum $\sum_{1}^{n_1} p_\nu$ is less than a. Since the sum $\sum_{1}^{n_1} p_\nu$ increases with n_1 beyond all bounds, it is always possible, by using enough terms, to make the partial sum larger than a.

The sum will then differ from the exact value a by at most p_n. We now add just enough negative terms $-\sum_{1}^{m_1} p_\nu$ in order to ensure that the sum $-\sum_{1}^{m_1} p_\nu \; -\sum_{1}^{m_1} q_\nu$ is less than a. This is also possible, as follows from the divergence of the series from the divergence of the series $\sum_{1}^{\infty} q_\nu$.

The difference between this sum and a is now at most q_{m1}. We now add just enough other positive terms $\sum_{n_1+1}^{n_2} p_\nu$ in order to make the partial sum again larger than a, as is again possible since the series of positive terms diverges. The difference between the partial sum and a is now at most p_{ms}.

We again add just just enough negative terms $\sum_{m_1+1}^{m_2} q_\nu$, which start next after the last one previously used, to make the sum once more less than a and continue in the same way.

The values of the sums thus obtained will oscillate about the number a and, when the process is carried out far enough, the oscillation will only take place between arbitrarily narrow bounds; in fact, since the terms p_n and q_n themselves tend to 0 when n is sufficiently large, the length of the interval in which the oscillation occurs will also tend to 0, and the theorem is proved.

In the same manner, we can rearrange s series in such a way as to make it diverge; we merely have to choose such large numbers of positive terms as compared with the negative ones that there occurs no longer compensation.

Operations with Infinite Series: It is clear that two convergent infinite series $a_1 + a_2 + ... = S$ and $b_1 + b_2 + ... = T$ can be added term by term, i.e., that the series formed from the terms $c_n = a_n + b_n$ converges and has the sum $S + T$, because

$$\sum_{\nu=1}^{n} c_\nu = \sum_{\nu=1}^{n} a_\nu + \sum_{\nu=1}^{n} b_\nu \rightarrow S + T$$

This theorem is really nothing more than another statement of the fact that the limit of the sum of two terms is the sum of their limits. It is also clear that, if we multiply each term of a convergent infinite series by the same factor, the series remains convergent and its sum is multiplied by the same factor. In the cases just mentioned, it is immaterial whether the convergence is absolute or conditional. On the other hand, a further study, which is not necessary for us here, shows that, if two infinite series are multiplied by each other by the method used in multiplying finite sums, the product series will not usually converge or have the product of the two sums as its sum unless at least one of the two series is absolutely convergent.

Tests for Convergence and Divergence

We have already encountered a test of a general nature for the convergence of series, which applies to series with terms of alternating signs and decreasing absolute values and which asserts that such series are at least conditionally convergent. In what follows, we shall only consider criteria referring to absolute convergence.

The Comparison Tests

All such considerations of convergence depend on the comparison of the series in question with a second series; this second series is chosen in such a way that its convergence can readily be tested.

The general comparison test may be stated as follows:

If all the numbers $b_1, b_2 ...$ are positive and the series $\sum_{\nu=1}^{\infty} b_\nu$

converges and if $|a_n| \leqq b_n$ for all values of n, then the series $\sum_{\nu=1}^{\infty} a_n$ is absolutely convergent. If we apply Cauchy's test, the proof becomes almost trivial, because, if m n, we have

$$|a_n + ... + a_m| \leqq |a_n| + ... + |a_m| \leqq b_n + ... + b_m .$$

Since the series $\sum_{\nu=1}^{\infty} b_\nu$ converges, the right hand side is arbitrarily small, provided that n and m are sufficiently large, whence for such values of n and m the left hand side is also arbitrarily small, so that, by Cauchy's test, the given series converges. The convergence is absolute, since our argument applies equally well to the convergence of series of absolute values $|a_n|$.

The analogous proof for the following fact will be left to the reader:

If $|a_n| \geqq b_n > 0$ and the series, $\sum_{\nu=1}^{\infty} b_\nu$ diverges, then the series $\sum_{n=1}^{\infty} a_\nu$ is certainly not absolutely convergent. Comparison with the Geometric Series: In applications of the test, the most frequently used comparison series is the geometric series. We obtain immdediately the theorem:

The series $\sum_{n=1}^{\infty} a_n$ is absolutely convergent, if from a certain term onwards there holds a relation of the form,

$$|a_n| < cq^n,$$

where c is a positive number independent of n and q is any fixed positive number less than 1. This test is usually expressed in one of the following, weaker forms: The series,

$$\sum_{n=1}^{\infty} a_n$$

converges absolutely, if there holds from a certain term onwards a relation of the form

$$\left|\frac{a_{n+1}}{a_n}\right| < q,$$

where q is again a positive number less than 1 and independent of n, or, if from a certain term onward there holds a relation of the form

$$\sqrt[n]{|a_n|} < q,$$

where q is again a positive number less than 1. In particular, the conditions of these tests are satisfied if there applies a relation of the form

$$\lim_{n\to\infty}\left|\frac{a_{n+1}}{a_n}\right| = \mathrm{k} < 1$$

Let the criterion IIa, the ratio test, be satisfied from the suffix n_0 onward, i.e., when $n > n_0$. For the sake of brevity, let $a_{n_0+m+1} = b_m$ and find that,

$$|b_1| < q|b_0|,\ |b_2| < q|b_1| < q^2|b_0|,\ |b_3| < q, \text{ etc.;}$$

hence $|b_m| < q^m|b_0|$, which establi shes our statement. For the Criterion IIb - the root test -we have at once $|a_n| < q_n$, and our statement follows immediately. We consider an arbitrary number q such that $k < q < 1$. Then, from a certain n_0 onwards, i.e.,, when $n > n_0$, it is certain that,

$$\frac{a_{n+1}}{a_n}\Bigg| < q \text{ or } \sqrt[n]{|a_n|} < q,$$

as the case may be, since from a certain term onwards the values of,

$$\frac{a_{n+1}}{a_n}\Bigg| < q \ of \sqrt[n]{|a_n|}$$

differ from k by less than $(g - k)$. The statement is then established by reference to the results already proved. We stress the point that the four tests derived from the original criterion $|a_n| < cq^n$ are not equivalent to each other or to the original one, i.e., that they cannot be derived from one another in both directions. We shall soon see from examples that, if a series satisfies one of these conditions, it need not by any means satisfy all the other ones. None of these statements can he reversed. For the sake of completeness, it may be pointed out that a series certainly diverges if from a certain term onwards $|a_n| > c$ for a properly chosen positive number c or if from a certain term onwards $\sqrt[n]{|a_n|} > 1$, or if

$$\lim_{n\to\infty}\left|\frac{a_{n+1}}{a_n}\right| = k, \qquad \lim_{n\to\infty}\left|\sqrt[n]{a_n}\right| = k,$$

where $k > 1$. In fact, as we immediately recognize, in such a series the terms cannot tend to zero as n increases, whence the series must diverge. (Under these conditions, the series cannot even be conditionally convergent.) Our tests furnish sufficient conditions for the absolute convergence of a series, i.e., if they are satisfied, we can conclude that the series converges absolutely. However, they are definitely not necessary conditions, i.e., absolutely convergent series can be found which do not satisfy these conditions. For example, the knowledge that,

$$\lim_{n\to\infty}\left|\frac{a_{n+1}}{a_n}\right| = 1, \qquad \lim_{n\to\infty}\left|\sqrt[n]{a_n}\right| = 1,$$

does not entitle us to make any statement about the convergence of the series. Such a series may converge or diverge. For example, the series $\sum_{n=1}^{\infty}\frac{1}{n}$, for which $\lim_{n\to\infty}\sqrt[n]{|a_n|} = 1$ and $\lim_{n\to\infty}\sqrt[n]{\left|\frac{a_{n+1}}{a_n}\right|} = 1$ is divergent, as we have seen before. On the other hand, we shall soon see that the series $\sum_{n=1}^{\infty}\frac{1}{n^2}$, which satisfies the same relations, converges. As an example of the application of our tests, we first consider the series $q + 2q^2 + 3q^3 + ... + nq^n + ...$

For this series,

$$\lim_{n\to\infty} \sqrt[n]{|a_n|} = |q| \cdot \lim_{n\to\infty} \sqrt[n]{n} = |q|,$$

$$\lim_{n\to\infty} \left|\frac{a_{n+1}}{a_n}\right| = |q| \cdot \lim_{n\to\infty} \frac{n+1}{n} = |q|.$$

It follows from the ratio test as well as from the root test even in the weaker form III that the series converges, if $|q| < 1$.

On the other hand, if we consider the series $1 + 2q + q^2 + 2q^3 + ... + q^{2n} + 2q^{2n+1} + ...$, we can no longer prove convergence by the ratio test when ýÿ $\nmid$ $|q| < 1$, because then $|2q^{2n+1}/q^{2n}|$ $= 2|q| \nmid 1$. But the root test yields immediately $\lim \sqrt[n]{|a_n|} = |q|$ and shows that the series converges provided $|q| < 1$, which, of course, we could have observed directly.

Comparison with an Integral

We now proceed to a discussion of convergence which is independent of the preceding considerations. We shall carry it out for the particularly simple and important case of the series,

$$\sum_{n-1}^{\infty} \frac{1}{n^{\alpha}} = 1 + \frac{1}{2^{\alpha}} + \frac{1}{3^{\alpha}} + ...,$$

where the general term a_n is $1/n^a$, $\alpha > 0$. In order to investigate the properties of this series, we consider the graph of the function $y = 1/x\alpha$ and mark off on the x-axis the integral values $x = 1, x = 2$. We first construct the rectangle of height $1/n\alpha$ over the interval n-1 x-n of the x-axis ($n > 1$) and compare it with the area of the region bounded by the same interval of the x-axis, the ordinates at the ends and the curve $y=1/x\alpha$. Secondly, we construct the rectangle of height $1/n$a lying above the interval nx n+1 and similarly compare it with the area of the region lying above the same interval and below the curve.

In the first case, the area under the curve is obviously larger than the area of the rectangle; in the second case, it is less than the area of the rectangle.

In other words,

$$\int_{n}^{n+1} \frac{dx}{x^{\alpha}} < \frac{1}{n^{\alpha}} < \int_{n-1}^{n} \frac{dx}{x^{\alpha}},$$

as we may also prove directly from the integral itself. Writing down this inequality for $n = 2$, $n = 3$, ..., $n = m$ and summing, we obtain the estimate * for the m-th partial sum,

$$s_m = \sum_{n=1}^{m} \frac{1}{n^{\alpha}}$$

$$1 + \int_{2}^{m+1} \frac{dx}{x^{\alpha}} < s_m < 1 + \int_{1}^{m} \frac{dx}{x^{\alpha}}$$

From this relation follows at once that the sequence of numbers,

$$C_n = 1 + \frac{1}{2} + \frac{1}{3} + \dots + \frac{1}{n} -$$

log n is bounded below. Since we see from the inequality,

$$\frac{1}{n+1} < \int_{n}^{n+1} \frac{dx}{x} = \log (n+1)$$

that the sequence is monotonic decreasing, it must approach a limit,

$$\lim_{n \to \infty} C_n = \lim_{n \to \infty} \left(1 + \frac{1}{2} + \frac{1}{3} + \dots + \frac{1}{n} - \log n\right)$$

The number C with the value 0.577,2 is called *Euler's constant*. In contrast to the other important special numbers of analysis, such as p and e, no other expression with a simple law of formation has been found for Euler's constant.

Now, as m increases, the integral,

$$\int_{1}^{m} \frac{1}{x^{\alpha}} dx$$

tends to a finite limit or increases without limit according to whether $\alpha > 1$ or $\alpha > 1$. Consequently, the monotonic sequence of the numbers s_m is bounded or increases beyond all bounds according to whether $\alpha >$ or $\alpha > 1$, whence follows the theorem: The series,

$$\sum_{n=1}^{\infty} \frac{1}{n^{\alpha}} = \frac{1}{1^{\alpha}} + \frac{1}{2^{\alpha}} + \ldots$$

converges - and, of course, absolutely - if and only if, $\alpha > 1$. The divergence of the harmonic series, which we have proved previously in a different way, is an immediate consequence of this. In particular, we see that the series,

$$\frac{1}{1^2} + \frac{1}{2^2} + \frac{1}{3^2} + \ldots, \quad \frac{1}{1^3} + \frac{1}{2^3} + \frac{1}{3^3} + \ldots, \ldots \text{ converge.}$$

The series,

$$\sum_{\nu=1}^{\infty} \frac{1}{\nu^{\alpha}},,$$

the convergence of which we have just studied, frequently serve as comparison series in investigations of convergence. For example, we see at once that for $\alpha > 1$ the series $\sum_{\nu=1}^{\infty} \frac{c_\nu}{\nu^{\alpha}}$ converges absolutely if the absolute values $|c_n|$ of the coefficients remain less than a fixed bound independent of n.

Continuity of the Sum of a Uniformly Convergent Series

As we have already hinted, the significance of the uniform convergence of n infinite series lies in the fact that a uniformly convergent series in many respects behaves exactly like the sum of a finite number of functions. For example, the sum of a finite number of continuous functions is itself continuous and, correspondingly, we have the theorem:

If a series of continuous terms converges uniformly in an interval, its sum is also a continuous function. The proof is quite simple. We subdivide the series $f(x) = g_1(x) + g_2(x) + \ldots$ into the m-th partial sum $f_n(x)$ plus the remainder $R_n(x)$.

As usual,

$$f_n(x) = g_1(x) + \ldots\ g_n(x).$$

If now any positive number ε is assigned, we can,by virtue of the uniform convergence, choose the number n so large that the remainder is less than $\varepsilon/4$ throughout the entire interval, whence $\left|R_n(x+h)-R_n(x)\right|<\frac{\varepsilon}{2}$ for every pair of numbers x and $x+h$ in the interval. The partial sum $f_n(x)$ consists of the sum of a finite number of continuous functions, whence it is continuous; thus, for each point x in the interval, we can choose a positive δ so small that $\left|f_n(x+h)-f_n(x)\right|<\frac{\varepsilon}{2}$, provided $|h| < \delta$ and the points x and $m+h$ lie in the interval. It then follows that

$$\begin{aligned}\left|f(x+h)-f(x)\right| &= \left|f_n(x+h)-f_n(x)+R_n(x+h)-R_n(x)\right| \\ &\leqq \left|f_n(x+h)-f_n(x)\right|+\left|R_n(x+h)-R_n(x)\right|<\varepsilon,\end{aligned}$$

which expresses the continuity of our function. The significance of this theorem becomes clear when we recall that the sums of non-uniformly convergent series of continuous functions are not necessarily continuous, as our previous examples show. We may conclude from the preceding theorem that, if the sum of a convergent series of continuous functions has a point of discontinuity, then in every neighbourhood of this point the convergence is non-uniform, whence every representation of discontinuous functions by series of continuous functions is based on the use of non-uniformly convergent limiting processes.

Integration of Uniformly Convergent Series

A sum of a finite number of continuous functions can be integrated term by term, i.e., the integral of the sum can be found by integrating each term separately and adding the integrals. In the case of a convergent infinite series, the same procedure is possible, provided that the series converges uniformly in the interval of integration.

A series $\sum_{\nu=1}^{\infty} g_\nu(X) = f(X)$ which converges uniformly in an interval can be integrated term by term in that interval, or, more precisely, if a and x are two numbers in the interval of uniform convergence, the series,

$$\sum_{\nu=1}^{\infty}\int_a^X g_\nu(t)\,dt$$

converges and, in fact, converges uniformly with respect to x for each fixed value of a, its sum being equal to $\int_a^X f(t)\,dt.$.

In order to prove this we write, as before,

$$f(x) = \sum_{\nu=1}^{\infty} g_\nu(x) = f_n(x) + R_n(x)$$

We have assumed that the separate terms of the series are continuous, the sum is also continuous and therefore integrable. Now, if ε is any positive number, we can find a number N so large that, for every $n > N$, one has $|R_n(x)| < \varepsilon$ for every value of x in the interval. By the first mean value theorem of the integral calculus, we have,

$$\left|\int_a^x \{f(t) - f_n(t)\}\,dt\right| \leqq \in l,$$

where l is the length of the interval of integration. Since the integration of the finite sum $f_n(x)$ can be performed term by term, this yields,

$$\left|\int_a^x f(t)\,dt - \sum_1^n \int_a^x g_\nu(t)\,dt\right| \leqq \in l$$

However, since εl can be made as small as we please, this yields $\sum_{\nu=1}^{\infty}\int_a^x g_\nu(t)\,dt = \lim_{n\to\infty}\sum_{\nu=1}^{n}\int_a^x g_\nu(t)\,dt = \int_a^x f(t)\,dt$, as was to be proved. If we wish to conside instead of infinite series sequences of functions, our result can be expressed as follows: If in an interval the sequence of functions $f_1(x)$, $f_2(x)$, ... tends uniformly to the limit function $f(x)$, then $\int_a^b f(x)\,dx = \lim_{n\to\infty} \int_a^b f_n(x)\,dx$ for every pair of numbers a and b in the interval; in other words, we can then interchange the order of the operations of integration and going to the limit.

This fact is far from being a triviality. It is true that, from a naive point of view such as prevailed in the Eighteenth Century, the interchangeabi]ity of the two processes is hardly to be doubted; however, shows that in the case of non-uniform convergence the above equation might not hold. We need only consider Example 2, in which the integral of the limit function is 0, while the integral of the function $f_n(x)$ over the interval $0 < x < 1$, i.e., the area of the triangle has the value $\int_a^b f_n(x)\,dx = n^{\alpha-2}$, and, when $\alpha > 2$, this does not tend to zero.

We see here immediately from figure that the reason for the difference between $\int_0^1 f_n(x)\,dx = n^{\alpha-2}$, and $\lim_{n\to\infty}\int_0^1 f_n(x)\,dx$ lies in the non-uniformity of the convergence.

On the other hand, by considering values of α in the interval $1 < \alpha < 2$, we see that the equation $\lim_{n\to\infty}\int_0^1 f_n(x)\,dx = \int_0^1 f(x)\,dx$ can be valid, although the convergence is non-uniform.

As a further example, we can integrate term by term the series $\sum_0^\infty g_n(x)$,, where $g_n(x) = x^n - x^{n-1}$ for $n > 1$ and $g_0(x) = 1$, between 0 and 1, even though it does not converge uniformly.

Thus, while uniformity of convergence is a sufficient condition for term-by-term integrability, it is by no means a necessary condition. Neglect of this aspect may easily lead to a misunderstanding.

Differentiation of Infinite Series

The behaviour of uniformly converging series or sequences with respect to differentiation is quite different from that with respect to integration. For example, the sequence of functions $f_n(x) = \dfrac{\sin n^2\pi}{n}$ certainly converges uniformly to the limit function $f(x) = 0$, but the derivative $f_n'(x) = n\cos n^2 x$ does not converge everywhere to the derivative of the limit function f '(0) = 0, as we see by considering $x = 0$.

In spite of the uniformity of the convergence, we cannot therefore change the order of the process of differentiation and pass on to the limit. Naturally, corresponding statements hold for infinite series. For example, the series,

$$\sin x+\frac{\sin 2^4 x}{2^2}+\frac{\sin 3^4 x}{3^2}+\ldots$$

is absolutely and uniformly convergent, because its terms are numerically not greater than the terms of the convergent series,

$$\frac{1}{1^2}+\frac{1}{2^2}+\frac{1}{3^2}+\ldots.$$

However, if we differentiate the series term by term, we obtain the series $\cos x + 2^2 \cos 2^4x + 3^2 \cos 3^4x + \ldots$, which obviously does not converge everywhere, since, for example, it diverges at $x= 0$. The only useful criterion which assures us in special cases that term-by-term differentiation is permissible is given by the theorem:

If, on differentiating a convergent infinite series,

$$\sum_{\nu=0}^{\infty} G_\nu(X) = F(X)$$

term by term, we obtain a uniformly convergent series of continuous terms,

$$\sum_{\nu=0}^{\infty} g_\nu(X) = f(X),$$

then the sum of this last series is equal to the derivative of the sum of the first series. Hence, this theorem expressly requires that, after differentiating the series term by term, we must still investigate whether the result of the differentiation is a uniformly convergent series or not. The proof of the theorem is almost trivial, because, we can integrate term by term the series obtained by differentiation. Recalling that $g_n(t) = G_n'(t)$, we obtain

$$\int_a^x f(t)\,dt = \int_a^x\left(\sum_{\nu=0}^{\infty} g_\nu(t)\right)dt =$$

$$\sum_{\nu=0}^{\infty}\int_a^x g_\nu(t)\,dt = \sum_{\nu=0}^{\infty}\left(G_\nu(x)-G_\nu(a)\right) = F(x) - F(a)$$

Since this is true for every value of x in the interval of uniform convergence, it follows that $f(x) = f(a)$ as was to be proved.

Power Series

Among infinite series, power series occupy the main position. We mean by a power series a series of the type

$$P(x) = c_0 + c_1x + c_2x^2 + ... = \sum_{\nu=0}^{\infty} c_\nu x^\nu$$

(a power series in $\boldsymbol{x}$) or, more generally,

$$P(x) = c_0 + c_1x(x - x_0) + c_2(x - x_0)^2 + ... = \sum_{\nu=0}^{\infty} c_\nu \left(x - x_0\right)^\nu$$

(a power series in $(x - x_0)$), where x_0 is a fixed number. If we introduce into the last series $\xi = x - x_0$ as a new variable, it becomes a power series $\sum_{\nu=0}^{\infty} c_\nu \xi^\nu$ in the new variable ξ, whence, without any loss of generality, we can confine our attention to power series of the more special form $\sum_{\nu=0}^{\infty} c_\nu x^\nu$

We have considered the approximate representation of functions by polynomials and were thus led to the expansion of functions in Taylor series, which, in fact, are power series. We shall now study power series in somewhat greater detail and obtain expansions of the most important functions in series in simpler and more convenient ways than before.

Convergence Properties of Power Series

There are power series which converge for no value of x, except, of course, for $x = 0$, e.g., the series $x + 2^2x^2 + 3^3x^3 + ... + n^nx^n + ...$ In fact, if $x < 0$, we can find an integer N such that $|x| > 1/N$. Then the absolute values of all the terms n^nx^n, for which $n > N$, will be larger than 1 and, as n increases, nn^nx^n will increase beyond all bounds, so that the series fails to converge. On the other hand, there are series which converge for every value of x; for example, the power series for the exponential function,

$$e^x = 1 + x + \frac{x^2}{2!} + \frac{x^3}{3!} + ...,$$

the convergence of which for every value of x follows at once from the ratio test. The $(n + 1)$-th term, divided by the n-th term, yields x/n and, whatever number x is chosen, this ratio tends to zero as n increases. The behaviour of power series with regard to convergence is expressed in the fundamental theorem: If a power series in x converges for a value $x = \xi$, it converges absolutely for every value x such that $|x| < |\xi|$ and the convergence is uniform in every interval $|x|$ ⪯ η, where η is any positive number less than $|\xi|$. Hence h may lie as close to $|\xi|$ as we please.

The proof is simple. If the series $\sum_{\nu=0}^{\infty} c_\nu \xi^\nu$ converges, its terms tend to 0 as n increases. From this follows the weaker statement that all the terms lie below a bound M independent of n, i.e., $|c_n \xi^\nu| < M$. If now q is any number such that $0<q<1$ and we restrict x to the interval $|x|$ ⪯ $q|\xi|$, then $|c_n x^\nu| < |c_n x^\nu| q^\nu < Mq^\nu$.

Hence, in this interval, the terms of the series are smaller in absolute value than the terms of the convergent geometric series ΣMq^ν, whence follows, by the theorem, the absolute and uniform convergence of the series in the interval $-q|\xi| < x < q|x|$. If a power series does not converge everywhere, i.e., if there is a value $x = x$ for which it diverges, it must diverge for every value of x such that $|x| > |\xi|$. In fact, if it were convergent for such a value of x, by the theorem above, it would have to converge for any numerically smaller value of ξ.

Hence, we recognize that a power series, which converges for at least one value of x other than 0 and which diverges for at least one value of x, has an interval of convergence, i.e., there exists a definite positive number r such that the series diverges for $|x| > \rho$ and converges for $|x| < \rho$. No general statement can be made for $|x| = \rho$.

The limiting cases, in which the series converges only for $x=0$ and that which converges everywhere, are expressed symbolically by writing $\rho = 0$ and $\rho =$ ⪯, respectively. It is possible to find this interval of convergence directly from the coefficients c_ν of the series. If the limit $\lim_{n\to\infty} \sqrt[n]{|c_n|}$ exists, then

$$\rho = \frac{1}{\lim\limits_{n\to\infty} \sqrt[n]{|c_n|}}$$

In general, ρ is given by the formula

$$\rho = \frac{1}{\overline{\lim\limits_{n\to\infty}} \sqrt[n]{|c_n|}}$$

where $\overline{\lim}$ is the symbol for the upper limit.

For example, $\rho = 1$ for the geometric series $1 + x + x + \ldots$; at the end-points of the interval of convergence, the series diverges. Similarly, for the series for the inverse tangent,

$$\text{arc tan } x = x - x^3/3 + x^5/5 - + \ldots,$$

we have $\rho = 1$ and at both of the interval's end-points $x = 1$ the series converges, as we recognize at once from Leibnitz's test.

We deduce from the uniform convergence the important fact that within its interval of convergence (if such an interval exists) a power series represents a continuous function. Integration and Differentiation of Power Series: On account of the uniformity of convergence, it is always permissible to integrate a power series

$$f(x) = \sum_{\nu=0}^{\infty} c_\nu x^\nu$$

term by term over any closed interval lying entirely within the interval of convergence. We thus obtain the function

$$F(x) = c + \sum_{\nu=0}^{\infty} \frac{c_\nu}{\nu+1} x^{\nu+1},$$

for which

$$F'(x) = f(x)$$

Moreover, since $|c_{\nu/}(n+1)| \leq |c_\nu|$ for all values of ν, the series obtained by integration converges more rapidly than the original series. We can also differentiate a power series term by term within its interval of convergence, thus obtaining the equation,

$$f'(x) = \sum_{\nu=1}^{\infty} \nu c_\nu x^{\nu-1}$$

In order to prove this statement, we need only show that the series on the right hand side converges uniformly, if x is restricted to an interval lying entirely within the interval of convergence.

Now, let x be a number, lying as close to r as we please,

for which

$$\sum_{\nu=1}^{\infty} c_\nu \xi^\nu$$

converges; then, as we have seen before, all the numbers $| c_{nn} |$ lie below a bound M independently of ν, so that,

$$\left| c_\nu \xi^{\nu-1} \right| < \frac{\mathrm{M}}{|\xi|} = N$$

Now let q be any number such that $0 < q < 1$; if we restrict x to the interval $| x | \nleq q | \xi |$, the terms of the series under discussion are not larger than those of the series,

$$\sum_{\nu=1}^{\infty} \left| \nu c_\nu q^{\nu-1} \xi^{\nu-1} \right|,$$

and therefore less than those of the series,

$$\sum_{\nu=1} N\nu q^{\nu-1}$$

However, in this last series, the ratio of the $(n + 1)$-th term to the n-th term is $\{(n+1)/n\}q$, which tends to q as n increases.

Since $0 < q < 1$, it follows by Criterion IIIa that this series converges.

Hence the series, obtained by differentiation, converges uniformly and by the theorem at the end of last section represents the derivative $f'_n(x)$ of the function $f(x)$, which proves our statement.

If we apply this result again to the power series,

$$f'(x) = \sum_{\nu=1}^{\infty} \nu c_\nu x^{\nu-1},$$

we find on differentiating term by term that,

$$f''(x) = \sum_{\nu=2}^{\infty} \nu(\nu-1) c_\nu x^{\nu-2}$$

and, by continuing the process, we arrive at the theorem: Every function represented by a power series can be differentiated as often as we please within the interval of convergence and the differentiation can be performed term by term.

As an explicit expression for the k-th derivative, we obtain,

$$f^{(k)}(x) = \sum_{\nu=k}^{\infty} \nu(\nu-1)\ldots(\nu-k+1) c_\nu x^{\nu-k},$$

or in a slightly different form

$$\frac{f^{(k)}(x)}{k!} = \sum_{\nu=k}^{\infty} \binom{\nu}{k} c_\nu x^{\nu-k} = \sum_{\nu=0}^{\infty} \binom{k+\nu}{k} c_{k+\nu} x^{\nu}$$

These two formulae are frequently useful.

Bibliography

Apostol, Tom M.: *Calculus: One-Variable Calculus with an Introduction to Linear Algebra,* Wiley, NY, 1997.

Boole, George: *A Treatise On The Calculus of Finite Differences*, Macmillan and Company, New York, 2000.

Boyer, Carl B.: *The history of the Calculus and its Conceptual Development*, Dover Publications, New York, 2009.

Courant, Richard: *Differential and Integral Calculus. Vol. II*, Wiley Classics Library, New York: John Wiley & Sons, 2018.

________; John, Fritz: *Introduction to Calculus and Analysis Volume 1*, Classics in Mathematics, Berlin, New York:Springer-Verlag, 2019.

Deborah, Hughes-Hallett: *Multivariable Calculus*, Wiley, New York, 2016.

Eisenbud, David; Harris, Joe: *The Geometry of Schemes*, Springer-Verlag, 2018.

Gerald, B. Folland: *Real Analysis: Modern Techniques and Their Applications,* John Wiley & Sons, NY, 2004.

Goursat, Édouard: *A Course in Mathematical Analysis: Vol 1: Derivatives and Differentials, Definite Integrals, Expansion in Series, Applications to Geometry*, Dover Publications, New York, 1999.

Hardy, Godfrey Harold: *A Course of Pure Mathematics*, Cambridge University Press, 1998.

Hildebrand, F. B.: *Finite-Difference Equations and Simulations*, Prentice-Hall, Englewood Cliffs, New Jersey, 2008.

Hille, Einar; Phillips, Ralph S.: *Functional Analysis and Semi-groups*, American Mathematical Society, Providence, R.I., 2014.

Jackson, J.D.: *Classical Electrodynamics*, New York: John Wiley & Sons, 2018.

Kline, Morris: *Calculus: An intuitive and physical approach*, John Wiley and Sons, NY, 2007.

Kock, Anders: *Synthetic Differential Geometry,* Cambridge University Press, Cambridge, 2016.

Larson, Ron; Hostetler, Robert P.; and Edwards, Bruce H.: *Calculus of a Single Variable: Early Transcendental Functions*, Houghton Mifflin Company, 2013.

Lieb, Elliott; Loss, Michael: *Analysis*, AMS Chelsea, 2011.

Marsden, J.E.: Vector Calculus, W. H. Freeman & Company, 1996.

Michael J. Crowe: *A History of Vector Analysis: The Evolution of the Idea of a Vectorial System*, Dover Publications; New York, 1997.

Morris, Kline: *Mathematical thought from Ancient to Modern Times,* Oxford University Press, 2010.

Phillips, Ralph S. and Hille, Einar: *Functional Analysis and Semi-groups,* American Mathematical Society, Providence, R.I., 2014.

Richtmeyer, D. and Morton, K.W.: *Difference Methods for Initial Value Problems*, Wiley, New York, 2007.

Schey, H. M.: *Div Grad Curl and all that: An Informal Text on Vector Calculus,* W. W. Norton & Company, UK, 2015.

Sutherland, W. A.: *Introduction to Metric and Topological Spaces,* Oxford University Press, Oxford, 2017.

Index

N

O

P

Q

R

S

T

U

V

□□□